The Knox Mine Disaster
January 22, 1959

THE FINAL YEARS OF THE
NORTHERN ANTHRACITE INDUSTRY
AND THE EFFORT TO REBUILD
A REGIONAL ECONOMY

Robert P. Wolensky
University of Wisconsin-Stevens Point

Kenneth C. Wolensky
Pennsylvania Historical and Museum Commission

Nicole H. Wolensky
Marquette University

Commonwealth of Pennsylvania
Pennsylvania Historical and Museum Commission
1999

Commonwealth of Pennsylvania

Tom Ridge, Governor

Pennsylvania Historical and Museum Commission

Janet S. Klein, Chairman

James M. Adovasio

William A. Cornell

Thomas C. Corrigan, Representative

Andrea F. Fitting

Edwin G. Holl, Senator

Nancy D. Kolb

John W. Lawrence, M.D.

Stephen Maitland, Representative

Brian C. Mitchell

LeRoy Patrick

Allyson Y. Schwartz, Senator

Allen M. Wenger

Eugene W. Hickok, ex officio,
Secretary of Education

Brent D. Glass, Executive Director

Copyright © 1999 Commonwealth of Pennsylvania
ISBN 0-89271-081-0
Second Printing

Dedication

To the anthracite mineworkers who harvested the fuel that fired the American industrial revolution, including the nearly 35,000 who perished in mining accidents, and especially
the twelve who died in
the Knox Mine Disaster
January 22, 1959
and
to our father, and grandfather, Nicholas Wolensky (1917-1971), the son of Eastern European immigrants, who worked for over twenty years at the Harry E. Coal Company breaker in Swoyersville, Pennsylvania.

Second Anniversary

Caroline Baloga (wife of Knox Mine Disaster victim, John Baloga) and children.

The twenty-second is today
And we still cry and pray;
The children long for you
And God only knows I do too.
No matter what I say or do
I just can't forget you;
For the things we did and shared alike,
I just don't know if I'm doing what's right.
When I needed help I called to you
And you always came to my rescue.

But now it's not the same,
Because you're not here when I call your name;
A home that was so happy and true
Is filled with memories of you.
I try to smile but instead I weep
Because you're under ground so deep;
A mine disaster we can't forget
Since your body is down there yet.

You're needed and wanted, Dear Dad,
Here at home it's very sad;
God took you away from us so fast
And we asked Him why couldn't it last?
I Guess God has strange ways of doing things
That we don't know;
And somehow he picked you
And you had to go.

How much you suffered we will never know;
And God gave us a tremendous cross to bear,
For some good reason we must always wear.

God took away the only one we love
But our prayers will go to him up above,
God gave us happiness and took it away
And to Him we will always pray.
Yes we know we will meet him some day,
But the twenty-second is the day
Our Dad from us was taken away;
The year was nineteen fifty-nine,
And for us two years is a long time.

Published in the *Sunday Independent*,
January 22, 1961.

MINE DISASTER

Rena Baldricia, 1993

One score and four,
The Knox Mine Disaster
Was permitted
On the Devil's floor.
This tragic day
Forever to remember;
Centuries away.
But God intervened,
Seventy miners he saved
From a dark water's grave

From company hands,
False order they relay
For miners to dig
Black diamond coal
No matter how or where.
Time not be delayed.
For profit was the goal.
But God intervened,

Seventy miners he saved
From a dark water's grave.

What heartbreak and folly
It was to convey
That twelve coal miners'
Bodies just floated away.
Their comrades alive
Scrambled and clawed
To reach the light of day.
But God intervened,
Seventy miners he saved
from a dark water's grave.
Forever interred
Twelve souls roam
Haunting each passageway.
Their buddies they search;
Wherever they may lay.
But God intervened,
Seventy miners he saved
from a dark water's grave.

No miner will dig coal,
For under water is their claim.
The families of all
Survive with pain.
In sadness and prayer
As they light a candle
Beneath their loved one's name.
But God intervened,
Seventy miners he saved
From a dark water's grave.

Table of Contents

Acknowledgments ix

Introduction xv

Chapter 1. The Susquehanna River is Breached
January 22, 1959 1

Chapter 2. Plugging the Breach and Sealing the River Slope 45

Chapter 3. What Does it Mean? How Could it Have
Happened? 65

Chapter 4. Establishing Blame and Responsibility 83

Chapter 5. Rebuilding the Regional Economy in the
Post-Anthracite Era 125

Appendix I Oral History Interviews 147

Appendix II Glossary of Anthracite Mining Terms 153

Index 159

Acknowledgments

Numerous individuals and organizations assisted, directly and indirectly, with the research and publication of this study. At the University of Wisconsin-Stevens Point, the University Personnel Development Committee granted generous funding assistance, and the Department of Sociology as well as the Center for the Small City offered a supportive research environment. Special thanks to E. Sherwood Bishop, chair of Sociology; Edward J. Miller, chair of Political Science and co-director of the Center for the Small City; and Justus F. Paul, dean of the College of Letters and Sciences for their long-standing encouragement of the project. Miller, along with William Skelton, professor of history, and Robert Enright, professor of sociology, critiqued the text and have given much appreciated collegiality on this and other projects. Chris Christensen, Angela Gonzalez, Amanda Stack, and Anne Inghram provided important research assistance.

Robert Wolensky spent two semesters and several shorter visits at the Institute for Research in the Humanities, University of Wisconsin-Madison, where the first three chapters of this book were drafted. Many thanks to Paul Boyer, director, Loretta Frieling, program assistant, and the Institute faculty.

Several individuals at The Pennsylvania Historical and Museum Commission provided invaluable assistance with the project. We would like to thank Brent D. Glass, executive director; Frank Suran, director of the Bureau of Archives and History; Robert Weible, chief of the Division of History; and Linda Ries of the Division of Archives and Manuscripts for their support and encouragement. Linda Shopes, who directs the Scholars-in-Residence Program, was an early supporter. Robert Wolensky participated in this program in 1995, which allowed him to review the Pennsylvania Coal Company papers housed at the PHMC's State Archives in Harrisburg. Diane Reed, chief of the Division of Publications and Susan Gahres, who was responsible for the lay-out and design of this book, devoted their excellent professional abilities in carrying the project through to completion. Thanks also to Michael O'Malley for editing and Kim Krammes for photo editing a version of the Knox story "Disaster–or Murder?–in the Mines" which appeared in the Spring 1998 edition of *Pennsylvania Heritage*.

At the Wyoming Historical and Geological Society, Michael Bertheaud, executive director; Jessie Teitelbaum, librarian/archivist; and Ruth Bevan, library assistant, provided valuable support during numerous research excursions to the Wilkes-Barre area.

Wilkes University offered library, office, and other services on numerous occasions. Special thanks to Professors Phil Tuhy, Tom Bigler, and Brad Kenney; former Wilkes administrator, Ed Seminski; and to Wilkes' President Christopher Briseth for their hospitality and intellectual companionship.

Chester Kulesa, curator at the Anthracite Heritage Museum in Scranton, graciously furnished research materials and arranged for some significant oral history interviews. We have also enjoyed working with Chester and Steve Ling, the museum's site administrator, on an exhibit to commemorate the fortieth anniversary of the Knox disaster, which will run through 1999.

At the Penn State Labor History Archives, Denise Conklin made the path through the UMWA papers all the easier thanks to her professionalism and genuine interest in labor history. Staff members at the National Archives in College Park, Maryland, assisted with the study of federal government documents relevant to the disaster.

George Harvan, probably the most important living anthracite photographer, offered higher education on anthracite history and graciously permitted us to use his Knox photographs in this book. Lance Metz of the National Canal Museum in Easton afforded a forum for our early Knox research at the museum's annual symposium.

The staff of Earth Conservancy, Inc., including Rick Ruggiero, Tom Thomas, Michael Dziak, Ellen Alaimo, Pat Filipowich, Anna May Hirko, and Amy Gruzesky assisted with numerous research questions regarding the Glen Alden, Hudson, Lehigh and Wilkes-Barre, and Blue Coal Companies.

Diane Suffrin, executive director, and other staff members at the Osterhaut Free Library in Wilkes-Barre, including Elaine Sofranko and Maria Reno, provided substantial library and reference information.

Five local newspapers—*The Citizens' Voice, Scranton Times, Times Leader, Sunday Dispatch,* and *Catholic Light*—permitted access to their libraries and gave permission to use photographs. Special thanks to Allison Walzer, Bob Burke, Joe Dowd, John Watson, Jim Gittens, Paul Golias, and Jerry Zufelt.

Andrew Shaw, Min L. Matheson, Judge Max Rosenn, Judge Gifford Cappellini, Phil Tuhy, Ellis Roberts, Bill Hastie, Tom Bigler, Ed Schechter, Joe Keating, and Eric McKeever have served as incredibly consequential mentors in our ongoing study of anthracite history and culture.

We would like to express our deepest appreciation to the numerous individuals who gave their time and knowledge in oral history interviews. Our special thanks goes to the members of the Altieri, Baloga, Burns, Ostrowski, Sinclair, Featherman, Gizenski, and Stefanides families, who lost loved ones in the Knox Mine Disaster. All oral history participants are listed in Appendix I. Anthracite industry workers Joe Stella, Bill Hastie, George Gushanas, Frank Danna, and Frank Handley not only offered oral histories but were especially helpful in educating us about the industry as well as the disaster, while always being available to answer innumerable questions. Bill Hastie also read the entire manuscript and offered valuable suggestions.

Ron Slusser, Robert Janosov, and Tom Dublin provided crucial insights into the organization of the book. Ray Stroik, Charles Petrillo, Lynn Kincaid, Pat Hagan, and Jim Kelly contributed suggestions to early drafts. Robert Zieger provided intellectual encouragement and some important bibliographic sources.

The research would not have been possible without the tremendous support system and the locally relevant information provided by a large number of people in northeastern Pennsylvania and elsewhere: Ruth and Jack Hagan (Robert's in-laws who provided a home a way from home, as well as invaluable help with local history and people), Paul and Lillian Sheehan, Joe Costa (Costanzio Lopez), Sheldon Spear, and Edward J. Davies. Too numerous to mention individually, but invaluable nonetheless, were our relatives in the Carey, Dodelin, Hagan, June, Siracuse, Stroud, and Wolensky families. Thanks as well to Tony DeAngelo, John "Pluggy" Piazza, Wally and Pippi Wolinsky, Joe Cicero, Frank Roche, and Judy Kelly.

Most notably, we would like to thank our families. Robert would like to thank his wife Molly, now a full-fledged Wisconsinite, but born and raised in Kingston, who has provided oceans of love, support, and editorial assistance while patiently enduring weeks and occasionally months of her husband's absence (at which time something usually broke around the house!). Meredith Hagan Wolensky, also a Wisconsinite but very appreciative of her anthracite roots, critiqued the first three chapters, traveled numerous times to northeastern Pennsylvania with her dad and sister, and even assisted with a few interviews.

Kenneth would like to acknowledge Abby and Aaron Wolensky who, though Harrisburgers for all of their lives, are growing to understand their roots in the anthracite region and the commitment of their father, Ken, to the history of Pennsylvania and its people and industries. Ken's wife Cherie has supported this and related research and, like him, has learned much along the way. He is deeply indebted to her for years of encouragement.

Nicole would like to express sincere gratitude to her father, Robert, and uncle, Ken. They have given her the opportunity to not only learn about but actively participate in the recounting of an event prominent to the industry in which her family's roots lie. They have shared with her one of their deepest passions and in doing so have formed a bridge connecting a generation said not to value heritage with an important piece of the past. She would also like to thank her mother Molly, sister Meredith, grandparents Rose, Ruth, and Jack, and aunts and uncles for their immeasurable amounts of love, happiness, and lessons in life (as well as numerous trips in her younger days to memorable places like Angela, Hershey, Knoebels, Dorney, and Great Adventure parks!).

Our mother and grandmother, Rosalie Siracuse Wolensky, took care of us during our many visits home. She also provided volumes of information about her personal life story as well as our extended family's history which, together, constituted a microcosm of anthracite and Pennsylvania working class history. She enjoyed having her sons and grandchildren home. Our siblings, Carol and Jack Wolensky, also provided their support and encouragement, connected us to many people who provided relevant information, and also made us feel at home.

These individuals have tried to support and educate us. However, we take full responsibility for the contents of this volume and any errors that it may contain.

INTRODUCTION

In casual and serious conversations with some of our anthracite-minded friends—mineworkers and scholars alike—we have often heard them say that the full story about hard coal's demise in the northernmost of the four anthracite fields has yet to be told. In this book we propose to tell that story.

Many, if not most, observers mistakenly believe that the Knox Mine Disaster, suddenly and by itself, caused the death of the industry that powered the American industrial revolution. We have learned that the story is much more complicated than that. In fact, virtually all of the water that flowed underground as a result of the Knox disaster was pumped out during the months following. The catastrophe could more accurately be called the penultimate or nearly-fatal blow to a business that still constituted a large share of the area's economy despite four decades of steady decline. The cataclysmic events of January 22, 1959 did terminate mining in the middle portion of the field around Pittston, in the Wyoming Basin. But they did not affect mines to the north in the Scranton area because the Lackawanna Basin constitutes a separate "bowl" of anthracite. Deep mining continued throughout the 1960s and in to the early 1970s in parts of the Wyoming and Lackawanna sections. But if the Knox calamity did not kill the deep mining industry, what did?

Scholars have written about some of the causes including the competition from natural gas and oil and the high cost of pumping ordinary mine drainage. However, they have paid much less attention to other important factors such as inadequate capital investments by the large companies in mining and burning technologies, corporate and union deceit, illegal mining, the role of organized crime, and—perhaps most importantly—the contract-leasing system which is examined for the first time in this volume.

We have learned that the demise of anthracite is, in fact, a sordid tale involving greedy individuals, lax inspectors, and irresponsible corporations and unions. Indeed, the devastation that occurred in Port Griffith, Pennsylvania, resulted from a level of corruption that will still shock the public conscience. In chapter 4, we review the list of criminal indictments and trials that followed in the wake of the Knox disaster. These cases most likely revealed only a portion of the total malfeasance. For a culture of corruption had engulfed not only the Knox Coal Company but many other mining operations, as well as the United Mine Workers of America. These facts present an unfortunate but indisputable part of the story surrounding anthracite's final decades. We have found that many if not most older citizens in the northern anthracite field know about this culture of corruption because they have lived it. We suggest that younger citizens will need to learn more about it so they can deal effectively with the many social problems that the region faces and will continue to face in the twenty-first century.

Our first task in the pages that follow has been to document the events surrounding the break-in of the Susquehanna River at the Knox Coal Company's River Slope Mine. Second, we look at blame and responsibility. Third, we look at two explanations for the catastrophe—the "short answer" (focusing on individual negligence and greed) and the "long answer" (focusing on the "contract-leasing system" developed by the major coal companies in the decades preceding the disaster). Finally, we examine the public and private sectors' efforts to remake the post-anthracite economy.

One advantage of researching fairly recent events is that the people who experienced them can provide firsthand information. This study benefited immensely from over ninety oral history interviews contributed by members of the victims' families, mineworkers who escaped from the River Slope Mine, other Knox company employees, lawyers and a judge who participated in criminal trials, other anthracite employees, and ordinary citizens. The interviews are included in the authors' three-hundred-person Wyoming Valley Oral History Project housed at the University of Wisconsin-Stevens Point. As important as oral history has been, however, we also relied upon numerous coal company documents, newspaper accounts, court records, investigative hearings, official correspondence, United Mine Workers of America papers, and other written materials.

While the Knox tragedy should interest scholars concerned with anthracite culture and American industrial history, we are especially hopeful that citizens throughout Pennsylvania, and particularly in the anthracite region, will be drawn to the story. We believe that anthracite's people—local residents as well as members of the "diaspora"—should read about and relive the legend, not only to get an accurate sense of how it happened and what it meant, but to help them place the catastrophe in a larger context and thereby come to terms with this unfortunate aspect of the region's past. We further believe that the Knox disaster—and indeed the entire anthracite era—has bequeathed a regional legacy of personal problems and social injustices that persists to the present. Knowledge and understanding must be the among first steps in healing wounds, solving problems, and correcting injustices.

Chapter One
The Susquehanna River is Breached
January 22, 1959

The Susquehanna River flooded today and gouged a huge hole in a coal mine tunnel, spilling tons of water that trapped twelve. Forty-four miners were rescued.
   The New York Times, January 23, 1959[1]

Water poured down like Niagara Falls.
John Williams, assistant foreman, Knox Coal Company[2]

An Unexpected January Thaw

Unseasonable weather brought a fast thaw to eastern parts of the United States in late January 1959. Two days of sixty-degree temperatures coupled with unremitting rains led to road washouts, property damage, and flash floods. In northeastern Pennsylvania the weather turned the frozen Susquehanna River into a surging, ice-laden torrent. At the measuring station in Wilkes-Barre, the crest rose from 2.1 feet on January 20th, to just below the twenty-two foot flood stage on January 23rd.

Residents kept a close watch on the tide knowing that the Susquehanna had dispatched regular floods to communities in the surrounding valley, whose early Indian name, *Maughwauwame*, evolved into *Wiwaumic*, then *Wyomink*, and by the late eighteenth century, Wyoming. The valley is home to the cities of Nanticoke, Pittston, and Wilkes-Barre; the boroughs of Kingston, Swoyersville, Larksville, Plymouth, Edwardsville, Luzerne, Pringle, Forty Fort, Courtdale, Dupont, Ashley, Sugar Notch, Shickshinny, Laurel Run, Exeter, Warrior Run, West Wyoming, Wyoming, West Pittston, Laflin, Jenkins, and Yatesville; as well as the townships of Wilkes-Barre, Newport, Plymouth, Wright, Hunlock, Dallas, Hanover, Plains, and Jenkins (see fig. 1). Major floods inundated riverside municipalities in 1865, 1902, and 1936, while minor ones occurred virtually every year. The flooding problem abated considerably in the late 1930s when the Army Corps of Engineers built a modern levee system.

As the Susquehanna had usually preyed on homes, schools, and other surface structures, few could have anticipated the deluge that was about to strike the Knox Coal Company's River Slope Mine in Port Griffith, a small mining town near Pittston. And yet several of the Knox company's 174 employees were not completely surprised. They had expressed grave concerns about their workplaces under the Susquehanna. A few had experienced premonitions and nightmares, some complained to their superiors, others voiced fears and anxieties to relatives and friends.

Deplorably wet working conditions caused much of the concern. "Well, you used to have to work with a raincoat, and rain pants, and rain helmet on you," said laborer Joe Poluske. "If you would only stay there five minutes, you would be soaking wet."[3] The mine's location presented another problem. Lithuanian-born Mike Lucas, an imposing fifty-seven year old miner whose father had been killed in a mine accident fifty-four years earlier, nervously advised his foreman that "something's wrong" with the sharp upward path of the coal tunnel he was driving; was he getting too close to the river?[4] The foreman, Frank Handley, on the other hand, provided escape instructions to his crews and marked an exit route with chalk "just in case" an accident might occur.[5] Many workers knew they were quarrying precariously close to the surface because they could detect the rumble of the passing Lehigh Valley Railroad trains. "We could hear that diesel, too, when it went overhead," said Alex Kulikowich, a laborer.[6]

Gene Ostrowski, a "rockman" who dug tunnels through solid rock to

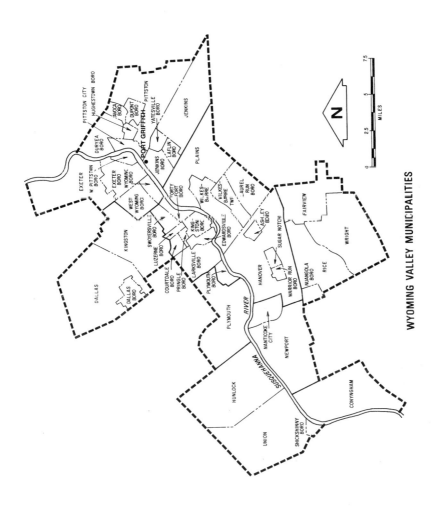

Fig. 1. The Wyoming Valley of northeastern Pennsylvania. (Courtesy of the City of Wilkes-Barre Planning Department)

access coal seams, had a disturbing premonition. It manifested itself in a recurring nightmare in which his bedroom ceiling cracked and collapsed on him and his young son as they lay sleeping in the same bed. In an uncanny replay of events, the "ceiling" in a River Slope tunnel did crack and sent the Susquehanna gushing into his electrician work area, killing the young father of three.[7] Company electrician Herman Zelonis expressed strong anxieties to his sister Veronica: "If that river comes in," he warned, "we'll be drowned like rats."[8] The river did come in, and Zelonis did drown.

In spite of these portents, most employees did not fully understand the risk-taking associated with their company's mining practices which were directed by President and General Manager Robert L. Dougherty— "The Bulldog" as the workers called him—and Secretary-Treasurer Louis Fabrizio. It proved to be a gamble that involved not only their jobs, but their very lives and the future of a still important industry. Indeed, it remains unclear whether the Knox bosses completely comprehended the risks they were taking. They did know that their enterprise paid handsome returns. They also realized that Knox reliably contributed to the financial balance sheet of the Pennsylvania Coal Company, from whom it leased the River Slope property and to whom it was required to sell all the coal it mined. They knew, finally, that their employees shared in the company's prosperity for, as long as they followed orders and met lofty production levels, they could count on steady, high-paid employment in a region where good jobs had become scarce.

In many ways the Knox Coal Company seemed an ideal small business. It earned millions of dollars and supported a well-paid work force. It became one of the more successful "new breed" of independent coal companies, those that, beginning in the 1920s, participated in the contract-leasing system fostered by the area's five major mining corporations. Under this arrangement, the five giants who owned or otherwise controlled the mineral rights—the Pennsylvania, Glen Alden, Hudson, Lehigh Valley, and Susquehanna coal companies—contracted and leased their mines to entrepreneurs and small producers, many of whom scavenged the coal, took chances with safety, cheated the employees, but nevertheless boosted output and profits in a declining industry. The contract-leasing system also helped corrupt District 1 of the United Mine Workers of America (UMWA) and, in the final analysis, facilitated the downfall of the once mighty northern anthracite industry, a subject further addressed in chapter 4.

Northern Anthracite Coal Field

The region's Native Americans, as well as the early British settlers, knew about the black rock that outcropped along the banks of the Susquehanna. The settlers called it stone coal because of its reluctance to burn. Later this premium grade of coal would be known as hard coal as opposed to its lower grade relative, bituminous, or soft coal.[9] By the late eighteenth century, blacksmiths had begun using anthracite to heat metal. Home and industrial uses began to grow after 1825, and within a few decades, anthracite had become the preferred fuel of a growing nation.

Northeastern Pennsylvania held 75 percent of the world's and 95 percent of the northern hemisphere's anthracite reserves. The hard coal district covers a relatively small domain—125 miles long and 35 miles wide. Although it spans a ten-county area of some thirty-three hundred square miles, only 484 square miles—a territory about the size of Manhattan Island—are underlain with workable seams. The region has been divided into four distinct fields: the southern, headquartered at Pottsville; the western-middle, located between Mahanoy City and Shamokin; the eastern-middle, centered around Hazleton; and the northern, situated in the Wilkes-Barre/Wyoming Valley and Scranton/Lackawanna Valley areas (see fig. 2).

The Knox Mine Disaster occurred in the northern field, the site of the region's first coal discovery in 1762. This field extends sixty-two miles from Forest City in Lackawanna County to the north, through the Wyoming Valley into Luzerne County in the middle, down to Shickshinny at the south. The field is split near the center by the Moosic Saddle, a geological divide that creates two rich "bowls" of anthracite which geologists have called the Wyoming and the Lackawanna Basins.

The northern field contained some of the region's deepest and highest quality coal. Due to its remote location, it was the last section developed. However, because its thick veins were pitched at relatively modest angles, the field delivered more efficient mining compared to the other districts, which had more steeply pitched seams lying in deep mountain valleys. Mining at great depths required considerable amounts of capital investment, so the northern field attracted the largest and most economically powerful corporations. The southern region dominated the industry during the first part of the nineteenth century, but by 1870 the north-

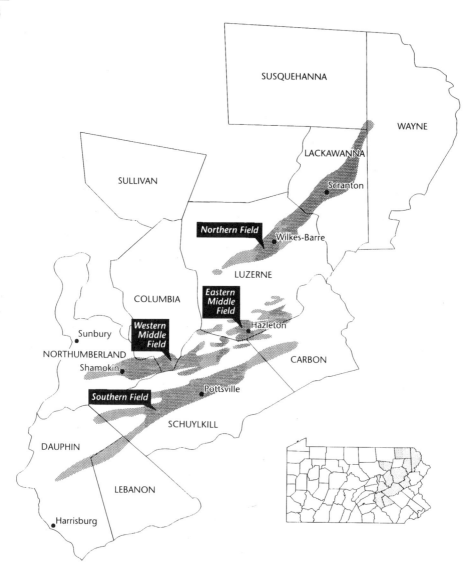

Fig. 2. The anthracite coal fields of northeastern Pennsylvania. (From E. Willard Miller, *A Geography of Pennsylvania*, [University Park, 1995] 216, courtesy of Penn State University Press)

ern area had become the largest producer of America's most important fuel. It emerged as the wealthiest and most financially influential of the four anthracite districts. By the last quarter of the nineteenth century, Scranton and Wilkes-Barre had become the region's dominant cities.[10]

The anthracite industry witnessed numerous accidents and calamities over the course of its history. The first known death occurred on February 23, 1823, when the roof inside a mine collapsed, burying a father and his young son. The three most deadly mishaps in the northern field took scores of lives: the Avondale Mine fire of 1869, where 110 men and boys were killed by fire and asphyxiation; the Twin Shaft cave-in of 1896, which cost seventy-five lives; and the Baltimore Tunnel explosion of 1919, where ninety-two mineworkers died.[11]

However, not since the mid-1880s had a major water-related mine accident hit the area. On December 18, 1885, twenty-six mineworkers died—their bodies were never recovered—in a water and quicksand inundation at the Susquehanna Coal Company's No. 1 Slope in Nanticoke, along the Susquehanna River. Thirteen years later, on May 8, 1898, an extensive cave-in saw the same river gush into the Pennsylvania Coal Company's Schooley Colliery in Exeter, just across the stream from the future site of the Knox Coal Company's River Slope Mine. No deaths occurred, but the Schooley Mine—which Knox later leased—suffered extensive damage and remained closed for nearly two decades.[12]

The calamity at the River Slope on January 22, 1959, though similar to the earlier incidents, was unique in one way. It represented the first and only instance where an inundation caused severe damage to several adjacent workings. As such, the Knox flood significantly contributed to the demise of northern anthracite, delivering a nearly fatal blow to an industry already in earnest retreat. The Knox catastrophe therefore arguably stands as the most destructive mine flood ever to strike the anthracite region. Its human death toll would have been among the worst if not for one of the largest and most dramatic escapes in American mining history.

THE RIVER SLOPE BREAK-IN

The River Slope operation began on May 26, 1954, when the Knox and Pennsylvania Coal Companies signed a lease that opened some of the finest coal in the district. The agreement gave the Knox access to the rich

Fig. 3. River Slope Mine entrance, circa 1955. (Courtesy of Jake Harenza, former Knox Coal Company employee)

virgin coal of the Pittston Vein, or the Big Vein as mineworkers called it, an eleven to nineteen-foot thick seam containing few impurities. The Pittston bed was the topmost of eight veins in the Port Griffith area. The coal had been part of a tract within the Pennsylvania Coal Company's Ewen Colliery, once a massive coal operation. With great expectations and enthusiasm, Knox officials contracted with the Gatti Engineering Company to dig the pit's entrance. Gatti completed the 230-foot slope entry, through solid sandstone, on July 24, 1954. The final cost came to $75,000. The mine became known as the River Slope because of its proximity to the east bank of the Susquehanna.[13]

The mine interconnected with several other Pennsylvania Coal Company properties such as the May and Hoyt shafts which were part of the Ewen Colliery; the Schooley Shaft across the river which was part of the Schooley Colliery; and an abandoned air shaft originally dug in 1868 by the long defunct Eagle Coal Company. The Knox Coal Company held leases on all of these areas. The linkages provided air passages as well as emergency exits.

At 7:00 A.M. on Thursday, January 22nd, the day of the unforget-

Fig. 4. Ewen Colliery, Pennsylvania Coal Company. (Courtesy of Joe Keating, Plymouth, Pa.)

table cataclysm, eighty-one workers reported for the first of three shifts. Seventy-five headed to their work places in the May Shaft, then the hub of mining activity, while another six traveled to the River Slope where the coal in Knox's first Pittston Vein lease had been exhausted. The latter group consisted of assistant foreman John Williams, laborers Fred Bohn and Frank Domoracki, who were removing mechanized equipment, as well as rockmen Gene Ostrowski, Charles Featherman, and Joseph "Tiny" Gizenski. The rockmen were actually employees of a tunnel contracting firm called Stuart Creasing, Inc. It was the last day of work for these men on a job to extend the slope into the second level vein, the Marcy, as part of a second lease Knox had secured from the Pennsylvania Coal Company.

Around 11:30 A.M. the laborers summoned Williams, a sixty-two year old Scottish-born mining veteran, to check the shrill cracking sounds coming from the timber ceiling supports, or props. "I no more than put my foot in the place and looked up," said Williams, in a characteristic burr, before a state investigative committee, "than the roof gave way. It

Fig. 5. Newspaper headline and story on the Knox Mine Disaster. (Courtesy of *Times Leader*)

sounded like thunder. Water poured down like Niagara Falls."[14] Unbelievably, the Susquehanna River had broken into the River Slope Mine.

Williams and the two laborers saved their lives by sprinting up the slope. Domoracki described the episode before state investigators:

Q. The roof just gave way?
A. And it gave way and the water come in right after it.
So Fred Bohn and Jack Williams, they turned—when they seen the water, they turned around and started running up the slope.
Q. Running up the slope to get out of the mine?
A. Out of the mines, that's right.
Q. What did you do?
A. I stood down there and I started hollering to the rockmen.
Q. Below?
A. Down below.
Q. That had been with Williams?
A. Yes. I hollered to Tiny [Joseph Gizenski], "Get out. The river broke in," and I started hollering down there until the water sealed the whole slope. You couldn't see no roof, no ribs sides of chambers], or. . . .

Q. When the water came in, did it go down the slope where those men were?

A. That's right, it sealed the whole place up. So when I seen the water sealed the whole slope, they couldn't get out, I turned around and I started running up the slope, and I passed Jack Williams and I just about got out when Fred Bohn got out and was leaning over the motor and I pressed him on the shoulder . . . and asked him, "How are you feeling, bud?" He told me, "If you can make it," he says, "go." So I ran in the engine house and I told the engineer—I'm winded, but I said, "Barney, call Bob Groves up and tell him to get everybody out of the May Shaft and shut off the power because the river broke in." And I says, "Tiny and his two buddies are drowned."[15]

Narrow Escapes and Tragic Deaths

Superintendent Robert Groves occupied an office in a shanty about a half mile from the stricken area. He immediately phoned underground and ordered a full evacuation. Believing it would prevent panic, he withheld the cause of the emergency.

Fear nevertheless spread among the men when they saw the surging water, heard its roar, and felt its power. According to assistant foreman Myron Thomas, the water

> sounded much like two speeding freight trains passing each other . . . [it] pushed my men all around the place. We grabbed on to old electric power wires which had long since been out of use to keep ourselves from being thrown off our feet."[16]

Laborer George "Bucky" Mazur stood in awe of the roaring underground rapids and its gushing icebergs "as big as desks."

Along with the three who ran up the slope, thirty-three others fled quickly on cages, or large elevator lifts, at the May and Hoyt Shafts. Mine Foreman Frank Handley retreated through the Hoyt opening with twelve others, the water surging to waist level before the cage began to move.[17] Tom Burns climbed on the elevator at this time.

> Frank Handley is one of the most courageous persons that I can lay it on to. He was waiting for us [in the mines] knowing at the time that the river was in, yet he waited at the

foot of this [inside] slope until all his men got accumulated. Then he told us that he was going to take us through the old workings to the Hoyt Shaft. . . .

Mr. Handley was ringing the bell for the cage to be lowered in the Hoyt Shaft. We stayed there about ten minutes and there was no answer to the bell. There were two bells to send the cage down, and the engineer up there [on the surface] gives you one bell noting that he's sending the cage down. Well, there was no answer.

Mr. Handley said we may as well go back and try to get up to the Baloga Slope. So we started and Mr. Handley still stayed at the cage, at the foot of the cage, and kept pressing that bell. He finally got an answer. That's when he called us back. By this time there was thirteen of us counting Mr. Handley. He called us back and the cage came down and we all got on the cage. Usually they only leave 10 people on the cage. Everybody crowded on the cage and Mr. Handley gave the signal. The water level was rising.

We got half way up, and the cage stopped! And then it started again. Well, we found out later—some of the things you find out later are really unbelievable. . . . [that] the rightful shaft engineer had the day off . . . [so] them cages were suppose to be out of commission. But Freddy Cecconi [an electrician] was sent over to run the engine that hoisted and lowered the cages. So he got over there and answered our bell. Well, he got our cage half way up in the shaft when the rightful engineer got there. Freddy had him take over to complete the operation of raising us out of the mines.

We didn't know that, of course, we just wondered what the hell he stopped for![18]

Mike Lucas headed for the May Shaft with his laborers Dan Stefanides and Willie Sinclair following. He waded into the gangway up to his neck until finally reaching the hoist. Turning for his fellow mineworkers and not seeing them, Lucas realized that their smaller statures had prevented them from breasting the powerful, bone-chilling tide. For days he worried about their survival.[19]

One of the most daring flights involved Ed Zakseski, a forty-six-

Fig. 6. Frank Handley, foreman, Knox Coal Company, 1988. (Photo by Robert P. Wolensky)

year-old former Army swim instructor. "Big Zack" quickly set out for the Hoyt Shaft with a few co-workers. "The waters rose like a tidal wave," he said, "It was choked with ice and timbers." He waded as long as possible then began swimming into the dark and treacherous chamber, dodging icebergs, timbers, and debris until, after about a half mile, the water level dropped. He came to a tightly closed door that served as an air seal. Through the portal lay the Hoyt cage. But where were his associates? They could not fight the torrent and were lost. Zakseski pried open the door and walked a short distance to the lift. He was one of the last persons rescued from this passage.[20]

At great risk to his own life, superintendent Groves climbed on to the cage with two assistants and descended into the May Shaft to search for survivors. However, the party could not venture any farther than the bottom because of the rapidly ascending waters. Indeed, the excursion almost led to another calamity. When the superintendent signaled the engineer to hoist the lift, it malfunctioned. The tide engulfed the would-be rescuers and Groves and his companions were forced to climb on top of the elevator. They narrowly avoided drowning before repairs were made and they were pulled to the surface.[21]

As dozens of other trapped workers scrambled from chamber to chamber searching for an outlet, company officials telephoned the deputy secretary of the Pennsylvania Department of Mines and Mineral Industries, Dan Connolly, who rushed to the scene from his home in Kingston. Connolly assumed leadership of the emergency, although federal mine inspectors eventually took control. At least five other state and five federal mine officials rushed to the scene. Several local, state, and federal elected officials arrived at the site, including the district's congressman, the flamboyant Daniel J. Flood.[22] A call went out over the airwaves requesting off-shift Knox miners to report in and assist with the rescue.

The immediate flight of thirty-six mineworkers meant that forty-five were still trapped inside. All would drown unless they could find a way out, or unless officials could quickly stopple the yawning cavity, 120-150 feet in diameter, that gouged the bed of the Susquehanna. The hole produced a massive whirling funnel along the river's east bank near the Lehigh Valley Railroad tracks. It drank millions of gallons of water and ate large chunks of ice, the latter going down "as if you put a cigarette butt in a toilet bowl and flushed the toilet"[23]

By mid-afternoon, hundreds of citizens gathered along the cliffs overlooking the river to watch the spectacle. Temperatures began to plummet as a blistering wind lashed out at rescuers as well as onlookers. State policemen arrived to provide safety and crowd control. As the afternoon wore on, hope for the missing men began to wane. Family members as well as officials feared that the forty-five mineworkers still unaccounted for had drowned. The area braced for what could be the most deadly mine accident in decades.

However, at about 2:45 P.M., a very wet and mud-covered laborer, Amadeo Pancotti, made a remarkably dangerous climb out of the earth through the abandoned Eagle air shaft. Located about three-hundred feet north of the breach along side of the railroad tracks, the air hole turned out to be the only escape hatch available. William Hastie, an off-shift company laborer charged with guarding an area around the site,

Fig. 7. Close-up view of the whirlpool in the Susquehanna River. (Courtesy of Wyoming Historical and Geological Society)

was the first person to meet Pancotti.

I was assigned to patrol the Lehigh Valley railroad tracks, which ran very close to the river in the bed of the old North Branch canal. I had orders to stop all traffic, rail or foot, because [of the] hole in the river bottom right against the bank. It was eating away the bank and it well could have eaten away under the railroad tracks.

Sometime later . . . on the way upriver I encountered Amadeo [Paul] Pancotti who was coming down the tracks. . . .

He was dressed in mining clothes and I thought he was a second shift miner waiting . . . off shift, really, because of the disaster, and satisfying his curiosity. I stopped him and I said, "Paul, I can't let you through because the tracks may cave in." He blurted something out very quickly which I didn't understand. Paul spoke in broken English. I began to scold him a little and he interrupted me, and this time with great vehemence, got it through to me that he had escaped from the mine!

He had escaped by way of an air shaft to the old, long

Fig. 8. William Hastie, Knox Coal company laborer (1989), and noted authority on the Knox Mine Disaster. (Photo by Robert P. Wolensky)

defunct, Eagle shaft mine. . . . He had climbed fifty-two feet up the sheer rock face [using] finger and toe holds. Now, probably the worst part of his climb was trying to get over the top because he'd only have loose dirt to clutch at. And if I'd just gone over there on my first impulse—because I had been in that area earlier and saw steam coming from the air shaft—I'd probably [have] been there to help him up over the edge. Well, I've kicked my seat ever since for not having done that. But at any rate, he told me that there were five more men at the foot of the shaft. I immediately called up to some men that were along the top of that cliff, night shift mineworkers, and yelled as loud as I could to get rope! Get rope! They ran off, scurried off in the direction of the of the engine house.

 I hurried over to that hole and I soon got company. They got rope to us in a hurry. It wasn't exactly rope, it was heavy electric cable with a heavy rubber covering but it sufficed. It turned out there were only [three] men at the foot of the shaft then. We got them out. We lowered the rope and got them out.[24]

The first refugee pulled from the cavern was Louis Randazza of Hughestown. A total of thirty-three men in two separate groups eventually climbed to the surface through this opening. They actually began their escape as one large assembly under the co-leadership of veteran mine employees Pacifico (Joe) Stella and Myron Thomas. The group splintered during the long trek to what the leaders had quickly realized was their only possible escape route. Pancotti traveled with a group of seven led by Stella, a thirty-five-year-old Pennsylvania Coal Company surveyor who happened to be at the site on January 22 conducting an inspection. Knox Assistant Foreman Myron Thomas led the twenty-five remaining men. According to Stella (JS):

JS: What happened when I was in there that day, I was making an inspection with Myron Thomas of what we called the substation area. . . . While I'm waiting for him, I was putting my map up to date of what we inspected in his working chambers. As he's having his sandwich we could hear this motor [a small electric locomotive] that transports the loaded cars to the shaft and would bring empties back for the men to load.

 Myron Thomas tells me, "I don't like the sound of that motor. It's really barreling in there. It [usually] doesn't come in that fast, and he's empty." So Myron was thinking that probably somebody was hurt or something see, and they called for the motor because we weren't in the office, we were out inspecting his section. The motor runner came right up to the door where we were standing. He said, "Myron, we just got word for everybody to get out of the mines." Myron told the motor runner, "You go back out and go down to Marcy and get those men out. Me and Joe will go back and get those guys out of the substation here."

Q: Did he say why you had to get out?

JS: No. He just said everybody get out of the mines as fast as you could. Back in we go, Myron and me. We get all the men. He had about three or four chambers working in there which would have amounted to about ten to twelve men. We got them all out and we met where they would go down to the Marcy. We waited there until this guy came up with the men from the Marcy. Then we said, "Okay, well then, let's start heading out."

We started to head out the main exit [May Shaft], the way we came in. I remember I was on the lead and I started stepping in water. You couldn't tell it was water because it was water that was rising slow and, in the mines, water that rises slow has a film of dust over the top of it [so] that you can't tell there's water there until you start stepping in it. I said, "Look, there's something wrong here." I'm stepping in water and as I'm going ahead the water's getting higher. I didn't go any further than about a foot high. My boots weren't too high. I said, "This is all water."

I wore a lamp with a spotlight. You could shine a distance. I'm shining down there and I could see where we had to go down the little grade and you could see where the water roofed, right to the roof. I said, "There's something wrong." We didn't know where the water was coming from. Myron and I decided the closest place would be to get out through the River Slope. I said, "Yeah, go ahead." We start heading toward the River Slope. As we're going toward the River Slope the pressure of the water was getting terrific. You couldn't get into the water. If you did you were gone.

The current was so swift. And the big chunks of ice! They tell me the ice that day was about sixteen or eighteen inches thick. Them chunks of ice was coming down and taking those mine cars and just pushing 'em. What saved us is that the workings were going up on a grade. You had the main heading, [it] was like in the basin and then they mined to the right and to the left which went uphill just like a "V". . . . What saved us is we could stay away from that fast water and cross from one chamber to the

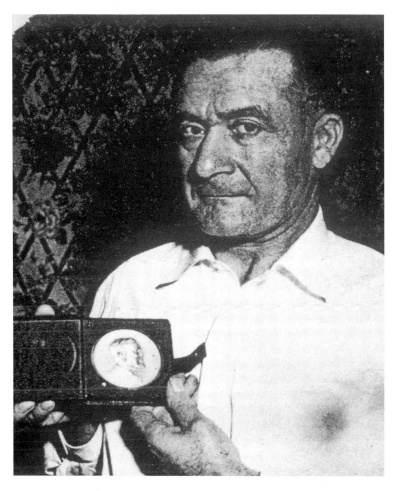

Fig. 9. Amadeo Pancotti holding his Carnegie Medal for Heroism. (Courtesy of Pittston *Sunday Dispatch*)

other by staying up in the higher part, up in the higher elevation.

We continued toward River Slope. . . . As we were getting closer to River Slope, boy the noise was so tremendous with the breaking of timbers and ice and everything. You couldn't hear yourself talk. You could only motion with your light, you know, wave to somebody if you want them to come over or either wave it across for somebody to stop. So finally, as we were getting closer to River Slope the chambers from this ten to fifteen degree angle, they started to level off down to ten, five [degrees], and then the water comes right up to the roof and we can't go any further. Myron and I decided the only chance we had was to head back to the old Eagle air shaft.[25]

In their march to the air shaft, six men fell to the rear. Stella, who carried detailed mine maps, remained with them. After a hazardous and wearisome excursion of more than two hours, the smaller group reached their destination only to find the air shaft's base clogged with tons of debris. Stella and two colleagues decided to travel back into the mines to find digging tools, while keeping watch for Thomas' main party. Pancotti and three others remained at the air shaft but grew impatient and decided to begin digging by hand. With persistence and care, they burrowed through some twenty vertical feet of rubble and emerged into the ten-foot wide passage. They bellowed for help but received no response. Pancotti then decided to attempt the precariously steep and icy fifty-foot ascent. The deed earned him the Carnegie Hero Fund Commission award as well as other honors.[26]

Meanwhile, Stella and his colleagues encountered swiftly rising waters that blocked their return to the Eagle Shaft. They strenuously worked around each obstruction and were continuing toward the shaft when they suddenly heard voices. To their great surprise they met foreman Frank Handley who was part of a rescue crew. Within a half hour, all members of the first group had been brought to the surface.

The joy over their deliverance quickly dissolved, however, as attention shifted to the remaining twenty-six hostages. Their situation had become desperate. Myron Thomas had a general idea of the air shaft's location but with most passages flooded and without adequate maps he could not find it. Laborer George "Bucky" Mazur described the predicament:

> We were trying to find the air hole, going all day. I'm talking walking, walking, what was it, a good six or seven hours? And every place we went we were up to, at least I was up to, my chest. Some of them were up to their neck in the water, trying to wade through it. There were chunks of ice big as desks that were hitting you. But we could never find the right place, we were always walking to the higher elevation to try to keep away from the water.[27]

The prospect of death prompted at least one of the men to prepare

Fig. 10. Joe Stella is rescued from the Eagle air shaft. (Courtesy of Pittston *Sunday Dispatch*)

for what might be the end. John "Stover" Gadomski had three brothers employed at the Knox, two were off shift and the other was George Mazur (GM), who described his half-brother's plan:

> GM: This is the only thing that shook me up. My brother took about three or four sticks of dynamite and caps and the plunger. He kept [them] with him. Then he said, "If we can't get out, we're not gonna be eaten up by rats." I said, "You gonna blow us up?" He said, "If we don't get out. You don't want to get eaten up by rats?" In other words, when we go to the high points and the water is coming, the rats are going to the high points, and if it gets to the end, you're not gonna drown. Just blow yourself up. . . .
> Q: So he was prepared to do that.
> GM: Yes he was. Yes he was.[28]

Gadomski freely admitted carrying the explosives to avoid a slow and painful death.

> I had dynamite. I told the guys—you ask my kid brother—I said, "I have dynamite." I said, "I am not drown-

Fig. 11. Joe Stella looking at the original River Slope map, 1988. (Photo by Robert P. Wolensky)

ing. If I have to put a wire in the battery, I will blow my head up before I go out and drown." Drowning is an awful death.[29]

At about 5:30 P.M., a second search party of several men descended into the Eagle air shaft. The explorers included Stella and Handley. They fanned out from the foot of the shaft keeping voice contact with the person to the right and left. Federal mine inspector Gerald Fortney and state mine inspector Warren Shirey remained at the top of the opening and directed efforts. Fortney eventually entered the mine and set off into some unexplored tunnels to search for Thomas' group. Myron Thomas recalled his desperate position:

> Leaving this one man about two hundred feet away from the main group, I told him to keep me in sight by watching

the reflector tape on the back of my helmet. About 150 feet from him, I noticed a large chamber with something white on the floor. Walking into it, I went up to my waist in water. The white was foam from the backlash of water which was building up.

Yelling back to the man on post, I told him it wasn't that way. Turning to my left, I noticed a small, rotten door in a wall. Going closer, I just could make out the letters "to Eg Shaft." I estimate the door to be forty or fifty years old.[30]

Shouting to the man on post that I had found the way and to bring the men up, I went through the door and found a large cave-in blocking our way. Examining it closer, I found there was a small arch with only enough room for us to crawl on our stomachs. On the other side, we found we were at the bottom of a rock tunnel slope.

The air was now blowing up the slope with great velocity. I noticed an argument among the men behind me. Going back, I asked what was wrong. One of the men wanted to turn the men back down the slope, telling them I was taking them into a trap. Telling them to settle down and listen to me, I told them if they would look back down the slope, they could see we had climbed about 100 feet. I explained that by now all other escapes were blocked with water and that the old Eagle air shaft was the only way out . . . That statement convinced them to follow me.

It wasn't but a little farther I thought I noticed a light ahead of me. Then it disappeared. Thinking I was seeing things, I did not say anything to the men about it. Later I found that Gerald Fortney, federal mine inspector, had a rescue crew with men posted at different intervals. He heard voices and going back to his men asked if they had talked. After they told him they had not, he returned to where he heard talking. I now saw and knew it was a fresh light. Yelling to Fortney, I went to him. He inquired who I was. After I introduced myself, he asked how many men I had with me. I told him between twenty and thirty. He told me to wait there and send up the men one at a time because there was a

8——THE SCRANTON TIMES, FRIDAY, JANUARY 23, 1959.

Led 30 Men to Safety

Mine Disaster Hero Tells His Story

Thomas Talks To Newsman

By MYRON THOMAS

(As told to Times Staff Writer Ed E. Rogers)

(Myron Thomas, 42, of 402 Union St., Taylor, the assistant mine foreman at the Knox Coal Co. operations in Port Griffith, led about 30 men to safety last night after the rampaging Susquehanna River broke through its banks and flooded the mine. This is his story as told to a Times reporter at Pittston Hospital).

We were working about 250 feet from the slope entrance when suddenly a motor runner and brakeman rushed in and told us to get everybody out of the mine right away. He didn't explain why.

I sent the motorman into the Pittston Vein to warn the men there while I hurried to the lower vein—the Marcey Vein— to warn my two crews (seven men) who were working there.

When I reached the Marcey I found the lights nad gone o at the foot, but one crew w out and I saw the lights of other men coming along gangway.

I strained my tonsils yel

MYRON THOMAS

Fig. 12. Myron Thomas, assistant foreman, Knox Coal Company. (Courtesy of *Scranton Times* and Robert Thomas)

bad roof at the surface shaft. Just ahead of us was an air shaft from [the] Pittston Vein to [the] Marcy Vein that was opened and unprotected.

A small ledge around the right hand rib [side wall] was the only way across the shaft. I stationed myself there because the men surged forward, knowing they were safe. I was afraid that they would fall down the shaft. I extended my hand to each man, thanking him for praying and asking God's help, and for his confidence in me.[31]

Steve Cigarski traveled with Thomas's group and managed to remain optimistic. "We never gave up hope. I knew there was an opening if we could only find it," he declared.[32] At 7:30 P.M., Thomas became the last mineworker rescued through an opening that, ironically, had been

Fig. 13. Miner is rescued from the Eagle air shaft. (Courtesy of Pittston *Sunday Dispatch*)

ordered sealed years earlier. "It was a state order," said Thomas, "Thank God someone didn't do it."[33]

Throughout the ordeal many relatives and friends kept a wearisome vigil near the air shaft, while others huddled in a small heated shanty near the main slope. According to one report:

> The cry of "They're bringing one out," signaled a rush from the building [shanty] and each rescued miner was surrounded by anxious relatives seeking word about the others.
>
> "Who is it?" one woman anxiously queried from the edge of the crowd, pushing her way to the rescued miner. Another grabbed him by the arm and asked if he had seen her husband. "I don't know anything. Let me through," the miner replied, the horror of his experience written on his face.
>
> One elderly miner cried like a baby when he was pulled to safety from the air shaft. Most said nothing. Trundled to the shanty by jeep over a railroad bed, the only way to get to the scene of the rescue shaft, the men were placed in ambulances at the shanty and then taken to the Pittston Hospital a

few blocks away.[34]

Bedlam characterized the situation at the hospital, as reported by one story:

> Corridors at the hospital began to fill with relatives and friends when word spread the miners had been rescued. Extra nurses, doctors and office staff worked frantically to give first aid treatment to the men and admit them as patients. Huddled in blankets, the exhausted, grimy miners were pushed in wheelchairs one by one to the hospital wards. Relatives walked with some, not wanting to leave the side of their loved one. Smiles of greeting were exchanged here and there by friends in the corridors.
>
> There was no happiness for some as they tearfully asked the information desk if their father or their husband were in the hospital, only to be told he was still missing. One student nurse, Theresa Orlowski, waited anxiously as each ambulance came in looking for her father, Frank Orlowski. She collapsed and was admitted to the hospital infirmary when she learned he was not among the rescued.[35]

Ed Borosky, one of those rescued, recalled his experience as a patient:

> They started stripping everybody because they were all soaking wet. I had a new pair of pants and a new pair of boots on. I only had them two days. They had to cut them [the boots] off me. I was all right. When I got to the hospital that was the first time I ever had a shot, a shot of whiskey. They poured it down my throat. They wanted us to stay there over night, but I had a young family at home. I wanted to go home. I got home in time to see my [TV] program [*Death Valley Days*] and my family.[36]

Following a jubilant reception, attention turned to the twelve still missing. Could any of them be found in time? Veteran miners thought that some might survive for a short while if they could find an air pocket, but without food and drinking water, their long-term chances were slim. In a region with strong religious traditions, the faithful participated in prayer services at the mine. One pastor, Fr. Edmund Langan of St. John's Church in Pittston, conducted a service and blessed the pit (fig. 14). Hope remained alive even after the rescue group returned to the surface

Fig. 14. Fr. Edmund Langan of St. John's Church in Pittston blessing the River Slope Mine. (Courtesy of *The Catholic Light*, Diocese of Scranton)

at 9:15 P.M. to avoid the surging water. When five SCUBA divers from the Civil Defense Underwater Rescue Team drove in from New Jersey and offered their services, authorities concluded that the subterranean current remained too strong and the tunnels too treacherous to permit their entry.

To the distress of twelve families, authorities decided to terminate the search. By 7:00 A.M. the next day, Friday January 23, water had completely filled the River Slope and engulfed the foot of the Eagle air shaft. Expectations dimmed further when, two days later, a guard reported methane gas leaking from the mine.

Meanwhile, television crews and news reporters began arriving from throughout the East. They transmitted footage and published photographs of the miners climbing out of the ground. They interviewed the rescued, the rescuers, and joyous and grieving family members. They spoke with local, state, and federal officials, including the recently inaugurated Pennsylvania Governor, David L. Lawrence. *The New York Times*, *Philadelphia Inquirer*, and other major metropolitan newspapers covered the story. *Life*

Fig. 15. Mike Lucas, Knox Coal Company miner, 1990. (Photo by Robert P. Wolensky)

magazine brought the Knox occurrence to national attention in the February 2, 1959 issue, with a brief article supplemented by three photographs–one a full-page aerial shot of the swirling Susquehanna vortex. On the same date, *Newsweek* published a detailed account accompanied by four photographs. The anthracite region and its deadly disaster had become the focus of national concern.

THE TWELVE VICTIMS OF THE KNOX MINE DISASTER

Some of the twelve who perished had briefly delayed leaving the mine, preventing them from reaching the available passages. Of course, they did not know the cause or the extent of the emergency. Sam Altieri received one of Superintendent Groves' phone calls and set out to notify

Fig. 16. Baloga family, photograph taken immediately after the disaster. From left to right: Audrey (age fifteen), John Jr. (six), Donald (standing, twenty-one), Sandra (thirteen), Mrs. Caroline. (Courtesy of Audrey Baloga Calvey)

others. Herman Zelonis decided to change his clothes. John Baloga wanted to put his tools away. Upon hearing the evacuation order, Willie Sinclair climbed up a long chamber to warn fellow crew members. By the time they all started toward the May Shaft, the water had reached hazardous levels so that neither Sinclair nor fellow laborer Dan Stefanides could follow Mike Lucas toward safety. Most likely the laborers ran away from the flow toward the chambers that branched out from the mine's main thoroughfare, or gangway. That area eventually flooded, however, and the duo perished. "To this day," said Lucas, "whenever I go near that river I spit into it, it bothers me so much."[37]

Along with the three rockmen who were certainly the first to drown, the survival possibilities of two others, Pennsylvania Coal Company employees Francis Burns and Benjamin Boyar, were nil. The lessor had the responsibility of maintaining the pumps so the two were repairing a siphon in one of the deepest veins, the Red Ash. But phone lines to the lower seams had not been maintained over the years so there was no way of warning them.

As with the Nanticoke inundation of 1885, all of the Knox victim remain entombed in the mine (see table 1). The ethnic diversity of the deceased—English, Irish, Italian, Lithuanian, Polish, Scottish, Slovak—reflected the historic immigration patterns of the anthracite region.[38] Herman Zelonis was the only unmarried victim; all of the others had families with children. Heartfelt compassion flowed to the grieving family members from a community that has called itself "The Valley With A Heart."[39] (see figs. 17-21)

Even before Amadeo Pancotti had emerged from the ground, however, state and federal mining officials began conferring with company personnel and local anthracite experts to develop a strategy for plugging the hole. At that time the lives of several employees seemed dependent upon some very expeditious decision-making. Moreover, as millions of gallons of water coursed underground nothing less than the future of numerous mining operations hung in the balance. Though their efforts took time and did not benefit the twelve who perished, the planners' often-revised strategy took full advantage of locally available resources.

Fig. 17. Stefanides family, photograph taken immediately after the disaster. From left to right: Patricia (age five), Daniel Jr. (nine), Mrs. Stephanie Stefanides, Michael (one), Christine (two). (Courtesy of Mrs. Stephanie Stefanides)

Fig. 18. Altieri family, photograph taken immediately after the disaster. From left to right: Mrs. Herman Ciampi (daughter), Sam Jr. and Vincent (sons), Mrs. Mary, Mrs. Nicholas Foglia (sister). (Courtesy of *Times Leader*)

Fig. 19. Ostrowski family, photograph taken immediately after the disaster. From left to right: Eugene (age twelve), Mrs. Theodosia, Anita (fifteen months), and Donna (five). (Courtesy of Ostrowski family)

Fig. 20. Daniel Stefanides holding Daniel Jr., circa 1952. (Courtesy of Mrs. Stephanie Stefanides)

Fig. 21. Eugene Ostrowski poses with son Eugene and daughter Donna during hunting season, 1957. (Courtesy of Ostrowski family)

Table 1[40]
The Twelve Victims of the Knox Mine Disaster

Victim	Age	Place of Residence	Occupation	Number in Family*	Years Mining Experience
Samuel Altieri**	62	Hughestown	Electrician	7	16
John Baloga**	54	Port Griffith	Miner	5	35
Benjamin Boyar***	55	Forty Fort	Electrical Foreman	3	28
Francis Burns***	62	Pittston	Lessor's Inspector	4	40
Charles Featherman****	37	Muhlenburg	Rockman/Laborer	2	12
Joseph Gizenski****	37	Hunlock Creek	Rockman/Foreman	4	22
Dominick Kaveliskie**	52	Pittston	Laborer	2	25
Frank Orlowski**	42	Dupont	Laborer	3	20
Eugene Ostrowski****	34	Wanamie	Rockman/Laborer	4	16
William Sinclair**	48	Pittston	Laborer	2	35
Daniel Stefanides**	33	Swoyersville	Laborer	5	2
Herman Zelonis**	58	Pittston	Electrician	1	40

*Immediate family, excluding the deceased; not all were dependents
**Employee of the Knox Coal Company
***Employee of the Pennsylvania Coal Company
****Employee of the Stuart Creasing Company, Rock Contractors

Notes to Chapter One

1. "River Flood Traps 12 Miners; 44 Rescued in Pennsylvania," *The New York Times*, January 23, 1959, 21.

2. John Williams, testimony before the Joint Legislative Committee to Investigate the Knox Mine Disaster, March 19, 1959, 893. Hereafter this source will be referred to simply as the Joint Legislative Committee.

3. Joseph Poluske, testimony, Joint Legislative Committee, March 12, 1959, 599-600. Poluske eventually quit working for the Knox company because of the intolerably wet and cold conditions. Assistant Foreman Myron Thomas was one of many who testified about the wet environment at the River Slope. See his testimony, Joint Legislative Committee, March 13, 1959, 789-790.

4. Michael Lucas, taped interview, June 25, 1990, Wyoming Valley Oral History Project (WVOHP), housed at the Center for the Small City, University of Wisconsin-Stevens Point, tape 1, side 1. Hereafter the oral history project will be referred to simply as WVOHP.

5. Foreman Frank Handley emphasized, however, he had no prior idea that a mine flood of this particular type and in this particular seam might occur. He was merely concerned with following the state mining law by telling his employees about a "second exit" escape route just in case any type of disaster occurred. He extensively marked the route to the Eagle air shaft with chalk. See Frank Handley, taped interviews, December 10, 1988 and June 27, 1990, WVOHP. As indicated in note number four here, Michael Lucas later relied upon Handley's escape instructions.

6. Alex Kulikowich, testimony, Joint Legislative Committee, March 19, 1959, 551; see also the testimony of Simon Zaboroski, March 19, 1959, 571-73.

7. Eugene Ostrowski, Anita Ostrowski Ogin, and Donna Ostrowski, taped interview, July 29, 1992, WVOHP, related this story about their father. If the disaster had occurred one week earlier, Mr. Ostrowski, would have avoided his demise. His work schedule changed from night to day shift just seven days before the disaster.

8. Herman Zelonis as quoted by his sister, Veronica, in David Morris, "Knox Disaster Killed Deep Mining Locally," *Times Leader*, January 22, 1983, 1A, 12A. Zelonis, the only unmarried victim, lived with his sister.

9. Geologists have classified coal into three general categories: lignite, bituminous, and anthracite (with grades in between). Anthracite is the "hardest" of the three; it contains the most carbon and burns the most efficiently. For a review of Pennsylvania's coal industry see E. Willard Miller, ed., *A Geography of Pennsylvania* (University Park: Penn State University Press, 1995), chapter 13.

10. The 1762 discovery of anthracite in the northern field is cited in Eliot Jones, *The Anthracite Coal Combination in the United States* (Cambridge: Harvard University Press, 1914), 7. On the historical development of the field see Howard Benjamin Powell, *Philadelphia's First Fuel Crisis: Jacob Cist and the Developing Market for Pennsylvania Anthracite* (University Park: Penn State University Press, 1978), especially chapters one and five; and Edward J. Davies II, *The Anthracite Aristocracy: Leadership and Social Change in the Hard Coal Regions of Northeastern Pennsylvania, 1800-1930*, (DeKalb: Northern Illinois University Press, 1985); Burton W. Folsom Jr., *Urban Capitalists: Entrepreneurs and City Growth in Pennsylvania's Lackawanna and Lehigh Regions, 1800-1920*, (Baltimore: Johns Hopkins University Press, 1981).

11. On the first mining death see Howard Benjamin Powell, *Philadelphia's First Fuel Crisis: Jacob Cist and the Developing Market for Pennsylvania Anthracite* (University Park: Penn State Press, 1978), 100-101. On the Avondale, Twin Shaft, and Baltimore Tunnel disasters see Ellis W. Roberts, *When the Breaker Whistle Blows: Mining Disasters and Labor Leaders in the Anthracite Region* (Scranton: Anthracite Museum Press, 1984). Of course, the northern field was not unique in this regard. All of the anthracite regions experienced accidents. For example, on the nineteenth century Schuylkill coal region, Anthony Wallace wrote, "Crushes, gas explosions, underground mine fires, and flooding (to put out the fires) put many collieries out of business repeatedly for weeks, months, or even years at a time." See Anthony F.C. Wallace, "The Miners of St. Clair: Family, Class, and Ethnicity in a Mining Town in Schuylkill County, 1850-1880, in David L. Salay, ed., Hard *Coal, Hard Times: Ethnicity and Labor in the Anthracite Region* (Scranton: Anthracite Museum Press, 1984), 11.

12. Other non-fatal water-related accidents in the Wyoming Valley include the Boroughs Shaft inundation of the Enterprise workings on July 4, 1872, when miners quarried into the bed of a canal; the Franklin

Mine flood of February 2, 1889, caused by a breach into a pond; the 1891 flood at No. 3 Colliery of the Susquehanna Coal Co. in West Nanticoke; the May 8, 1898 flood at the Pennsylvania Coal Company's Schooley Colliery in Exeter where a cave-in brought the Susquehanna into the mines; and the Midvale-Prospect Colliery inundation of December 19, 1915, when Mill Creek entered a chamber near an outcrop. See "Mine Floods Were Numerous Over the Years," *Sunday Independent*, January 25, 1959, sec. 1, p. 1, 5.

The 1898 Schooley Colliery deluge shut the mine down until it was pumped between 1913 and 1917. Then a rock tunnel was dug connecting it with the Pennsylvania Coal Company's Hoyt Shaft across the river. The Knox Coal Company's first lease with the Pennsylvania Coal Co. in 1943 was for the Schooley Shaft. The Schooley inundation is more fully discussed in, "River Came Through Once Before In Area Near Present Mine Flood Tunnel Across River Built in 1917," *Sunday Dispatch*, January 25, 1959, 35.

The *1961 Annual Report* of the Anthracite Coal Division, Pennsylvania Department of Mines and Mineral Industries, lists three other major anthracite flood accidents besides the Knox: February 4, 1891, Spring Mt. No. 1 Colliery of Spring Mt. Coal Company, Jeansville, Pa., thirteen drownings; April 20, 1892, Lytle Colliery of Lytle Coal Company, Minersville, Pa., ten drownings; and May 26, 1898, Kaska William Colliery, Dodson Coal Company of Middleport, Pa., six drownings (p. 45). A total of forty-six anthracite miners drowned in accidents between 1847-1963.

The Jeansville disaster occurred about thirty miles south of Wilkes-Barre, near Hazleton, and stands as perhaps the most dramatic mine flood preceding the Knox. Seventeen men were suddenly entombed when a dynamite charge broke into a body of water. All but four drowned. After eighteen days of pumping, the survivors were located. They were rescued on day twenty-two. Emaciated and near death, they had miraculously survived without light, water, or food, although some maintain that the four survived by eating rats' legs. On the Jeansville case see H.C. Bradley, ed., *History of Luzerne County, Pennsylvania* (Chicago: S.B. Nelson and Co. Publishers, 1893), 320; and Richard W. Funk, "Worst Mining Accident In Area Happened in Jeansville," *Standard-Speaker* (Hazleton), October 7, 1996, 17. The information on the survivors' eating rats' legs came from anthracite researcher and folklorist Joe Keating in a conversation on June 10, 1998.

One of the more notable inundations in the Lackawanna region occurred when the Lackawanna River, at flood stage, broke into the No. 10 Tunnel that operated under a lease to the Kehoe-Berge Coal Company from the Lehigh Valley Coal Company. The breach occurred only a few miles upstream from the Knox disaster.

The most recent mine deluge to strike the anthracite region occurred in 1977 at the Kocher Coal Company in Tower City, below Hazleton. Nine of ten trapped miners died. Associated Press writer Ronald Adley collected an oral history of the event from the lone survivor. Adley published a three-part account, released through the Associated Press, in regional newspapers on March 17, 18, and 19, 1977.

13. Gatti's difficult experiences in digging the River Slope entrance are detailed in his taped interview, July 14, 1995, WVOHP, tape 1, side 1.

There are four types of mine entrances. A shaft is a vertical entrance through which workers and coal are raised and lowered on two adjacent elevators (called cages). A drift is a surface-level mine that follows a seam into the side of a mountain. A tunnel mine is also at surface level but quarrying occurs up into a mountain containing coal seams. A slope is a twenty-five to thirty degree subsurface entry through which workers and coal travel on rail cars connected to a "rope" or steel cable.

14. "Water Poured Down Slope Like Niagara Falls, Mine Boss Says," *Times Leader, The Evening News,* February 17, 1959, 1; see also John Williams, testimony, Joint Legislative Committee, March 19, 1959, 893.

15. Frank Domoracki, testimony, Joint Legislative Committee, February 28, 1959, 140.

16. "24 Saved by Foreman, Say Surviving Miners: Taylor Man Led Group to Safety in 7 Hour Journey to Air Shaft," *Times-Leader, The Evening News,* January 23, 1959, 25.

17. Frank Handley, taped interview, December 10, 1988, WVOHP, tape 1, side 1. Handley provides the same story in "12 Missing and 35 Rescued After River Breaks Into Mine Near Pittston," *Wilkes-Barre Record,* January 23, 1959, 1.

18. Thomas Burns, taped interview, WVOHP, August 8, 1989, tape 1, side 1.

19. Michael Lucas, taped interview, June 25, 1990, WVOHP, tape 1,

side 1.

20. On Zakseski's retreat see "12 Miners Still Missing; Fire Occurs in Knox Slope: Rescue Work Halted As Water Continues to Pour into Mine," *Wilkes-Barre Record,* January 23, 1959, 1, 2; Libby Brennan, Tom Moran and Tom Heffernan, Jr., "Rays of Hope Found in Homes of 12 Men in Flooded Mines: Didn't Have a Chance," *Sunday Independent,* January 25, 1959, sec. 1, p. 1, 5. The WVOHP contains taped interviews with other men who quickly evacuated the mine: Chester Dunn, Michael Obsitos, William Hague, Joseph Kupcza, and Anthony Remus.

21. An account of Groves' nearly fatal trip can be found in William Hastie, taped interview, July 31, 1989, WVOHP, tape 1, side 1. Hastie, an employee at the Knox, was also Groves' son-in-law. The escape was also mentioned in George A. Sporher, "The Knox Mine Disaster: The Beginning of the End," *Proceedings of the Wyoming Historical and Geological Society,* 24 (1984): 141.

James Jamieson, an assistant foreman and one of Groves' companions on the journey, had minutes earlier escaped from the mine. A few hours later, he reentered the underground for a third time to assist with the search for stranded miners. See William Rachunis and Gerald W. Fortney, *Report of Major Mine Inundation Disaster, River Slope Mine,* 1959, 17.

22. Flood's first election to the Eleventh Congressional District occurred in 1944. He was defeated after his first term, reelected in 1948, defeated in 1952, then reelected in 1954, and served in Congress for fourteen terms. He became widely known as an effective politician who could deliver programs and favors for his constituents. In spite of his having resigned from the House in 1980 amid scandal, and subsequently pleading guilty on federal charges of accepting payments from individuals seeking government contracts, Flood's constituents remained very loyal. For more on Flood see George Crile, "The Best Congressman," *Harpers,* January 1975, 60-66; "Watch Out for Flood," *Newsweek,* February 13, 1978; "Flood: Dramatis Personnae," *People,* February 13, 1978; and William C. Kashatus III, "'Dapper Dan' Flood: Pennsylvania's Legendary Congressman," *Pennsylvania Heritage,* 21 (Summer 1995): 4-11.

23. George Gushanas, taped interview, January 13, 1994, WVOHP, tape 1, side 1. The diameter of the hole was variously reported as being between 75 and 150 feet. The likelihood is that it gradually grew as the gondolas and mine cars thrust into the hole to seal it actually reamed out

its sides. The lower figure was reported in "Pa. Mine Flood," *Coal Age*, 64 (March 1959), 28; while several newspaper reports gave the larger number.

24. William Hastie, taped interview, July 31, 1989, WVOHP, tape 1, side 2. Mr. Hastie contributed four other taped oral history interviews to the WVOHP on June 28, 1990; July 2, 1990; March 23, 1991; and March 13, 1996, and patiently offered numerous untaped conversations.

25. Joe Stella, taped interview, November 1, 1988, WVOHP, tape 1, side 1. In his testimony before state investigators, Stella agreed with Groves' decision not to inform the mineworkers that the river had broken in: "If word come in and said, 'Get out, the river was broken through,' I think none of us would have ever gotten out, because we would have gotten trapped like the rest of the other men on the way to the May Shaft." See Joe Stella, testimony, Joint Legislative Committee, March 20, 1959, 1056.

26. Mr. Pancotti was said to believe that he was not deserving of the medal and the honors which followed. See Peter Malak, "The History of the Anthracite Coal Industry Up to and Including the Knox Mine Disaster," bachelor's thesis, College Misericordia, April 1998, 21.

27. George Mazur, taped interview, December 20, 1988, WVOHP, tape 1, side 1.

28. George Mazur, taped interview, December 20, 1988, WVOHP, tape 1, side 1.

29. John Gadomski, taped interview, December 22, 1988, WVOHP, tape 1, side 2.

30. The marking was probably made by foreman Handley who had chalk-marked directions to the air shaft "on anything I could." Telephone conversation with Frank Handley, March 1, 1997.

31. Quote taken from an April 12, 1984, presentation by Myron Thomas at the Pennsylvania Anthracite Heritage Museum, Scranton. According to one report, Thomas did have some mine maps to facilitate his path finding. See "24 Saved by Foreman, Say Surviving Miners: Taylor Man Led Group to Safety in 7 Hour Journey to Air Shaft," *Times-Leader, The Evening News,* January 23, 1959, 25. Myron Thomas passed away in 1987, before the authors could include him in the WVOHP. We are grateful to Robert Thomas for providing written materials on his father as well as a taped interview of his own recollections, May 28, 1992, WVOHP.

32. "They Dug Way With Hands to Freedom," *Scranton Times*, January 23, 1959, 3.

33. "Knox Survivor: 'I Was Ready to Die;' Disaster Occurred 23 Years Ago," *Citizens Voice* (Wilkes-Barre), January 22, 1982, 4, 5.

34. "Relatives Huddle in Small Shanty to Await News of their Loved Ones," *Wilkes-Barre Record*, January 23, 1959, 17. The following evacuees received care at the Pittston Hospital the evening of January 22, 1959: John Pientka, Plains; Joe Stella, Pittston Township; Merle Ramage, Pittston; Joseph Solarczyk, Exeter; Joseph Francek, Pittston; George Mazur, Exeter; Louis Marsico, Old Forge; Anthony Krywicki, Port Griffith; Myron Thomas, Taylor; John Gadomski, Wyoming; Jerome Stuccio, Pittston; John Elko, Wyoming; James LaFratte, Pittston; John Piencotti, Exeter; Louis Randazza, Hughestown; Joseph Soltis, Port Griffith; Fred Cecconi, Plains Township; Angelo Rotendoro, Pittston; Michael Machachines, Inkerman; Joseph Kislevich, Pittston; John Balaro, Wyoming; George Shane, Inkerman; Charles Proshunia, Exeter; Charles Milchiulis, Pittston; John Gustitus, West Pittston; Stephen Cigarski, Port Griffith; Frank Wascalis, Exeter; Frank Ludzia, Pittston; John Moore, Sebastopol; Albert Smelster, Jenkins Township; Edward Borosky, Taylor; Joseph Kachinski, Wilkes-Barre; Martin Saporito, Pittston; Stanley Roman, Exeter; Paul Cawley, Port Blanchard. See "Names of Knox Men Treated at Hospital," *Times Leader, The Evening News,* January 23, 1959, 3.

35. "Relatives Huddle in Small Shanty to Await News of their Loved Ones," *Wilkes-Barre Record*, January 23, 1959, 17.

36. Ed Borosky, taped interview, August 4, 1989, WVOHP, tape 1, side 1.

37. Mike Lucas, conversation with Robert Wolensky, September 14, 1990.

38. Other examples of ethnic diversity could be seen in the clearly definable ethnic surnames of most Knox employees. Superintendent Robert Groves came from the same Scottish town as assistant foreman Jack Williams and victim William Sinclair. The company also employed Italian-born employees including Carnegie hero Amadeo Pancotti.

39. The authors would like to express their sincere gratitude to the following eighteen relatives of the deceased who participated in the Wyoming Valley Oral History Project: Francis Burns Jr. (son of Francis Burns Sr.); Anne Altieri Ferrare (daughter of Samuel Altieri), Frank Ferrare (son-in-law of Samuel Altieri), Yolanda Altieri Parenti (daughter of Samuel Altieri); Audrey Baloga Calvey (daughter of John Baloga), Donald Baloga

(son of John Baloga), Opal Featherman (wife of Charles Featherman); Ida Gizenski (wife of Joseph Gizenski), Al Gizenski (son of Joseph Gizenski), Joseph Gizenski Jr. (son of Joseph Gizenski Sr.); Jake Harenza (brother-in-law of John Baloga); Anita Ostrowski Ogin (daughter of Eugene Ostrowski), Donna Ostrowski (daughter of Eugene Ostrowski); Eugene Ostrowski, Jr. (son of Eugene Ostrowski Sr.); Lea Stark (wife of William Sinclair); Stephanie Stefanides (wife of Daniel Stefanides Sr.), Daniel Stefanides Jr. (son of Daniel Stefanides Sr.), Joseph Stefanides (brother of Daniel Stefanides). The dates of their interviews are listed in Appendix I.

A few members of the victims' families reported unusual occurrences associated with their fathers' deaths. Audrey Baloga Calvey said that the family's seven-day clock, wound by her father the day before he died, inexplicably ran for over three weeks after his disappearance. Yolanda Altieri Parente, who was out of the country at the time, had a dream about her father's demise the night before receiving a long distance phone call announcing the disaster. Donna Ostrowski, only a baby when her father passed away, had a dream a few years later when she was a young girl in which her father told her not to worry about him and not to miss him so much; when she awoke from the dream in the middle of the night the rocking chair next to the bed was rocking to and fro.

40. This table includes information gathered from personal interviews, newspaper reports, and from William Rachunis and Gerald W. Fortney's, *Report of Major Mine Inundation Disaster, River Slope Mine,* 1959, Appendix A. William Hastie was especially helpful in reconciling some differences between the sources.

Joseph Czarnecki's name easily could have appeared on this list, according to his wife, because one of the victims (she did not know which one) took his place on the January 22 morning shift. Emily Czarnecki, telephone conversation, July 25, 1993.

Chapter Two
Plugging the Breach and Sealing the River Slope

The Knox cave at the Knox Coal Company is without precedent as far as Army Engineers are concerned. We never had a case where the bottom fell out of a river.
    Captain Norman J. Drustrup,
    Navy Construction Battalion[1]

Plugging the Breach

Within hours of the catastrophe, federal and state mine officials began working on a plan to cork the massive whirlpool that curled along the banks of the Susquehanna River. The effort began when crews deposited dozens of smaller mine cars, tons of boulders, truck loads of coal waste, and hundreds of bales of excelsior wood shavings and hay into the breach. The fill arrived on huge dump trucks called Euclids that traveled up a makeshift, quarter-mile road over the railroad tracks from the south. A few local coal companies provided equipment, materials, and labor. However, after twenty-four hours of dumping, the river's flow had scarcely slowed. Much of the fill missed its mark and washed down river, while the mine cars were simply too small to fill the puncture. It became apparent that some alternate strategy was needed.

Advisors from the Army Corps of Engineers and the Navy Construction Battalion (Seabees) arrived to help revise the plan. To provide more bulk, several railroad coal hopper cars—behemoths called gondolas—each fifty feet long and weighing twenty-five tons, were shoved into the hole on the evening of Friday, January 23rd. Trainmen launched the gondolas from the Lehigh Valley Railroad route after work crews had installed a new set of tracks that branched off from the main road and curved toward the water's edge. Frank Danna, a superintendent for the Pagnotti coal enterprise, helped construct the new spur:

> What we did, we got trucks and we loaded [them with] railroad ties and rails. Then we built [new] railroad [tracks] from the old railroad [tracks] so they could put these railroad cars onto the track and run them right into the big hole.[2]

In a scene that persists as one of the most powerful visual legacies of the disaster, trainmen thrust one gondola after another into the massive hole using a yard locomotive (figs. 22-25). Arnold Embleton, a conductor on the Lehigh Valley train that day, remembered the formidable task:

> When I got to Coxton [Yard] the train master told me what we were going to do. He said, "You're gonna take," I think it was fifteen cars, he said, "We're gonna to push those down, out of here, to the Knox Mine Disaster." He said, "I will be with you." They picked old coal cars, hopper cars, that were in need of repairs but the air brakes on the cars were a hundred percent. We pushed the cars through Pittston. When we got down there we noticed that they had cut the track—this is the east-bound track—they had cut it and pushed it over towards the river. Realigned it to face the river. . . .
>
> They said that we would have to pull the train back towards Pittston and cut one car off [at a time] and close the knuckles and bleed the cars off [so] that it was free-wheeling. There would be no brakes on it. And then get a running start and push this car off the end of the track into the swirling

Fig. 22. Lehigh Valley Railroad tracks were cut and extended toward the river so that gondolas could be pushed into the whirlpool. (Courtesy of Anthracite Heritage Museum, Scranton)

hole in the river. By doing this we would get up a speed of maybe fifteen to twenty miles an hour. We'd flag the engineer and he would put the train into an emergency stop. One car would continue on off the end of the track.

The first two cars didn't go in the hole. The first two cars swirled around the hole and ended up down the river along the bank. They discovered that they would have to open the hoppers on these empty cars to allow the water to go in. The third car and the rest of the cars all went in the hole. They took us back to Coxton and they ordered another crew with the second train. They did the same thing as we did. . . .

There was nothing dangerous about it. In other words, the car was uncoupled, it was free wheeling and we had plenty of air in our train to keep us from going in, putting the air on the brakes. It seemed to work pretty good. As we got down to fewer cars, all the easier it was to stop the train.[3]

Despite the renewed effort, the opening still drank. Some people

Fig. 23. Gondola cars thrust into the Knox whirlpool. (Courtesy of Wyoming Historical and Geological Society)

argued that the gondolas actually made the hole larger by reaming its sides as they whirled around.[4]

George Gushanas (GG), superintendent for the Glen Alden Coal Company, was called in to supervise the campaign at the site of the breach.

GG: I looked at the dozer operator that was up there when the "EUCS" [dump trucks made by Euclid Road Machinery] would dump. He would push the load in. They didn't want to get them too close. It was Al Sorber. He used to work for me when I had charge at the strippings [open-pit mine] with the Glen Alden. He said to me, "Hi, George." I said, "Hi, Al." He said, "What ya' doing here?" I said, "The big boys sent me up to close this hole and I'm in charge now." [He said,] "Anything I can do to help?" I said, "Yes there is. There's a lot of help [needed]. Do you see where those trucks are backing in, EUCS, along the railroad track? Well right past there for another twenty to thirty feet is all them tele-

Fig. 24. Crews work to rebuild the washed-out river bank. (Courtesy of George Harvan)

graph poles and lines."

"Yeah, [he said], "that's why we can't do nothing." I said, "Well I'm telling you right now, you take your dozer over there and push all them poles and that copper wire and everything right in the river, all the way out where we got room to turn around. We want this road widened out at least two widths of a EUC." He laughed. He said, "I'll do it but I don't like doing it. You're gonna get in trouble." I said, "Don't worry about trouble. We got troubles!"

So he went and pushed all of them poles in. Then I went up and saw Pagnotti's man and I told him I wanted some finer breaker rock to put another road in there. The EUC start hauling the breaker rock down and we got a couple hundred loads of that down and we rolled it in. Al did it with his dozer. Then the EUCs would come in, go right to the hole, turn, back up and dump. . . . They had about a dozen EUCs working there when I got there. But after I widened the road out, we had about thirty to forty EUCS hauling.

. . . Then we decided we'd cut the railroad track, the Lehigh Valley railroad track, and well the "navvy gang" [track crew] was there that redid it, the cutting and the repairing, and they swung that track over to the hole. And then they went and they got us forty to fifty old dilapidated railroad cars, coal-hoppers. We'd take one of the cars, uncouple it, and close the coupling. The diesel engine would come up, give it a good shove, and it would go right in the river. That car would just keep going around a circle, down, down, and zoom she'd disappear. . . .

Q: How many cars went in the breach?

GG: Hundreds. And the railroad cars just about fifty. That was really something. . . .

Q: Did those mine cars and those railroad cars plug the hole up?

GG: The dirt itself plugged the hole up, the big rocks, and the stripping [materials] and all. Everything helped. We dumped in thousands of bales of straw and hay and excelsior. In bales now. We didn't tear it apart. Tractor trailers would come up with a whole load of bales. He'd open up the tailgate and just let them whole bunch of bales go down, hundreds of them. Inside of two-three seconds they were gone, sucked right down.[5]

On Saturday, officials decided to rebuild the washed-out river bank and construct a peninsular ramp out to the hole from which more fill, mine cars, and gondolas could be deposited. Euclids (dump trucks) continuously traversed the road until they dumped some 25,000 cubic yards of dirt, rock, boulders, and numerous small mine cars. A construction crane positioned near the water's edge hoisted more cars in to the puncture as well as a thirty foot by thirty foot steel "mat" fabricated by welding several railroad tracks into a cross-cut pattern. The mat, which actually broke up rather quickly once in the hole, was intended to settle on top of the materials already in the cavity and provide support for additional fill. About thirty more gondolas and four hundred smaller mine cars were tossed into the breach on Saturday. Finally, after two and one-half days of continuous effort, the gorge had been packed and the flow diminished to

Fig. 25. Onlookers watch as a gondola is tossed into the Susquehanna River. (Courtesy of Pennsylvania State Archives)

a relative trickle. A newspaper report described the last phase of the effort:

> The first encouraging sign to the men laboring to fill the hole in the river came about 6:30 last evening [Saturday].
>
> For the first time since the fight was started last Thursday some of the fill which had been dumped in failed to disappear.
>
> The heartening indication that perhaps the hole had been at last partially plugged raised spirits as the word passed.
>
> The good sign came after the road had been prepared over the breach and the big Cats and Euclids were able to run back and forth rapidly enough to pour in material in real quantities.

With that, one of the big railroad gondolas was pulled up and pushed in.

It rolled over, hit sideways but, luckily, filled with water just at the right time and went down like a stone.

Immediately, there was a noticeable lessening of the swirl in the whirlpool over the hole.

Another was brought up, went down sideways and the swirl eased still more.

When a third was brought up it was held until a decision could be made on the procedure to be followed since the loads of rock and fill poured in were for the first time staying, where they could be seen, giving rise to hope that at least some sort of a plug had been put in the hole.

A stop was called to dumping in the little mine cars, which today were going in four or five at a time and there must be 400 or 500 down there.

Then a meeting was cited to determine if this is the time to dump in still more gondolas or concentrate on rock fill.

The good news came about 10 o'clock last night after days of labor to stop the surge of roaring water into the mine cave at Port Griffith. . . .[6]

Laborers extended a peninsula out into the river to cover an area well beyond the chasm. However, twenty-four hours later a second, smaller void appeared, washing out a good portion of the fill. Workers mounted another fill-in maneuver. Three other smaller divots materialized over the next few days. In each case they were re-packed. At last, after one week, the area seemed secure.

Some 12.1 million gallons per minute, one-sixth of the river's total flow, entered the mines during the first forty-eight hours of the emergency.[7] Measuring instruments operated by the U.S. Geological Survey indicated that the underground water pool reached a peak of 502 feet above sea level at 8:00 P.M. on January 25th, significantly higher than the usual level of 100 feet.[8] Because of the Knox break-in, an estimated 10.37 billion gallons of water cascaded into the Wyoming Basin's mines.[9] The immediately inundated area extended about three and one-half miles by one-half mile.[10]

President Dwight D. Eisenhower sent a telegram to Governor

Fig. 26. Bales of hay arrive at the disaster site. (Courtesy of Stephen A. Teller)

Lawrence expressing his concern: "The distress suffered by the people in the stricken areas is of deep concern to me, and I hope that the situation may rapidly improve."[11] The president declared the site a disaster area and federal emergency funds soon followed. Governor Lawrence immediately convened a meeting of Luzerne County's seven state representatives and two senators to draft legislation to de-water the mines and permanently seal the affected area.

DE-WATERING THE MINES

Water had always presented serious problems for northern field mining. Indeed, deep mining was made possible only after 1850 when companies harnessed the steam engine to the mine pump. By one estimate, during the 1920s, the daily amount of so-called "make water"—ordinary mine drainage—pumped from anthracite mines surpassed the daily amount of water consumed in New York City.[12] For every ton of coal brought to the surface in the 1920s, companies removed some 10.5 tons of water. By the 1950s a ton of coal took forty tons of water. De-watering in the early 1950s cost mining corporations $12 million annually.[13]

Over the many decades, constant pumping kept the mine water at manageable levels.[14] When one firm went out of business a competitor or group of competitors assumed pumping responsibilities so their own mines would not flood. However, the Knox cataclysm required a pumping operation of a magnitude unseen in anthracite mining history.

Fortunately, most of the pumps, piping, and electrical equipment were at hand. Because the 1955 Federal-State Mine Drainage Program provided government subsidies for ordinary water evacuation, dozens of pumps had been ordered and were awaiting installation. However, several logistical problems had to be overcome before the equipment could become operational. Within one week of the disaster, crews were working on a twenty-four hour basis to install three different types of siphons. Labor, expertise, and equipment came from three local coal companies—Glen Alden Corporation, Hudson Coal Company, and Pagnotti Enterprises—plus the United States Steel Corporation of Pittsburgh, each providing a team of eight to ten workers.

Pump installation at the River Slope stalled for several days until hundreds of tons of ice chunks were removed from the mine. By late January, amid loud public complaints about the slow pace of emplacement, the first pump began withdrawing water from the River Slope. On February 3rd workers affixed another pump at the Schooley Shaft. These powerful devices withdrew 3,000 to 6,000 gallons per minute through twenty-four inch pipes.

As the underground pool levels declined, victims' families still hoped for a miracle. However, when Joseph T. Kennedy, secretary of the Department of Mines, announced on February 6th that at least three more weeks of de-watering would be needed before any underground search could begin, it finally became clear that the twelve were lost forever.

By mid-February, pumps at numerous locations were removing 45,500 gallons per minute and lowering the water level three feet every twenty-four hours. To facilitate flow back into the river, workers widened existing ditches and built several new ones. Installers placed the last deep well pump at the Hoyt Shaft on March 17. They also set sixteen more pumps in other

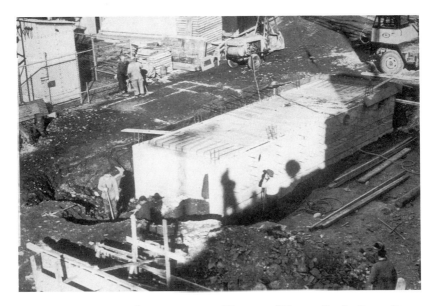

Fig. 27. Crews installing mine pumps. (Courtesy of Pittston *Sunday Dispatch*)

workings, bringing the grand total of siphons to forty. At their peak the pumps combined to discharge more than 142,000 gallons per minute. By early March, they had sufficiently lowered the pool in the River Slope so that the search for bodies and work on a permanent seal could begin.

SEALING THE BREACH AND SEARCHING FOR BODIES

In order to seal the hole and search for bodies the mine first had to be made safe for entry. The multifaceted engineering scheme began with the construction of an earthen cofferdam around the original breach, and the diversion of the Susquehanna.[15] The No. 1 Contracting Company, a subsidiary of the Pagnotti company, secured a $313,000 state contract for the task. It was the only firm in the area with sufficient equipment and expertise to handle the project, so it submitted the only bid. Commonwealth authorities wanted the job completed as soon as possible, so No. 1 Contracting employed 120 men divided across three seven-hour shifts.

In late March construction crews began building two roughly parallel dikes from the eastern bank of the river to Wintermoot Island near the opposite, western shore. They erected the third side of the cofferdam on

Fig. 28. Governor David L. Lawrence (right) is briefed by Secretary of Forests and Waters Maurice K. Goddard, and Secretary of Mines and Mineral Industries Joseph T. Kennedy (left) regarding the Knox Mine Disaster. (Courtesy of Pennsylvania State Archives)

the island. Meanwhile, the dam sent the river around the west side of the island into a pre-existing channel that had been dredged. This first stage of the project was completed on May 27th. Next, the crews pumped the water from inside the dam to expose the river bed. Trucks dumped tons of loam and clay over the actual breach to provide a stronger top seal.

At last, the time had come for an inspection team to enter the mine (fig. 29). The group entered through the Eagle air shaft. They examined the mine's condition and also searched for bodies. They undertook a second inspection a few days later, entering this time through the May Shaft. Knox Foreman Frank Handley (FH) participated in the excursions.

FH: After the water was pumped, the state Department of Mines had a commission of inspectors to inspect it. What I mean [is an] inspection and search [team] . . . Well, I was with the search team and [with] these inspectors. As you traveled, the mud was deep; sometime it'd get to your hips. . . . And you'd have sticks with a point on to probe ahead of you, looking for bodies . . .

Fig. 29. Newspaper story of the search and inspection crew. (Courtesy of *Sunday Independent*)

Q: Did you find any bodies?
FH: No.[16]

These journeys established the safety of the mine, allowing a fourteen-person workforce that included many former Knox employees to began work on a permanent underground seal. The workers reinforced the exact spot of the breach with iron bars to supplement the steel already contained in the crushed gondolas and mine cars. They built heavy-duty wooden bulkheads in two adjoining tunnels to keep the concrete in place. Then they poured over 1,200 cubic yards of concrete into the prepared area through a series of boreholes that had been drilled in the river's bed.

Frank Danna supervised this phase of the project:

> At that time I was superintendent of Franklin Colliery of No. 1 Contracting Company, a subsidiary of [Pagnotti's] Sullivan Trail Coal Company. [At first] I was ordered to shut down my operation and move men and material to help block the hole to stop the water. . . . [Then] I was appointed to do the mine work, and given complete charge. The Pennsylvania State Department of Mines and the U.S. Bureau of Mines were in charge of [planning] all operations. When the river was diverted, a pump was installed to pump the [remaining]

Fig. 30. Frank Danna underground at the River Slope viewing crushed gondolas and mine cars. (Courtesy of William Hastie)

water out of the mine. The slope area where the river came in was reopened, the heavy wooden stoppings were constructed at the bottom of each working place. A total of ninety bore holes—about nine inches in diameter—were drilled from the bottom of the river to the top of all working places. Tons and tons of concrete were poured down the bore holes and the mine was sealed. We took pictures of the most of the operation. The job was completed sometime in July 1959.[17]

The workers also constructed bulkheads in three other chambers that branched off from the River Slope and yet another in a hole in the slope's "floor" caused by the force of the original break-in. Nearly six hundred cubic yards of concrete were discharged into these sections. As an extra safety measure, supervisors decided to reinforce the areas in back of the concrete dams with over 26,000 cubic yards of sand flushed through twenty-six eight-inch boreholes.[18]

According to a report by federal mine inspectors, "This work reinforced the involved area sufficiently to stop pressure on pillars and tim-

Fig. 31. Patching the River Slope Mine. (Courtesy of William Hastie)

bers along and over the slope where the Pittston Vein was first intersected and where the break-in occurred."[19] Said Frank Handley, "I'll tell you, I think that seal could hold the Atlantic *and* Pacific Oceans."[20]

The final phase required the removal of the cofferdam, allowing the Susquehanna to resume its normal course. The state and federal governments assumed the full cost of the project with outlays of $2,100,000 and $1,249,394 respectively.[21]

Even before the breach was mended, however, citizens worried immensely about the economic and geological consequences of the calamity. They also asked the most immediately pressing question: how could it have happened? The next chapter examines the short-term effects as well as the causes of the disaster.

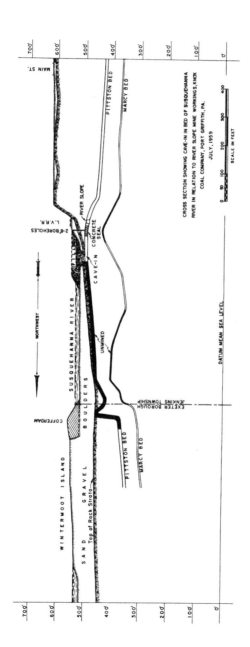

Fig. 32. Crossection of the River Slope Mine, showing the location of the concrete seal. [From William Rachunis and Gerald W. Fortney, "Report on Major Mine Inundation Disaster, River Slope Mine," U.S. Bureau of Mines (Wilkes Barre, 1959).]

Notes to Chapter Two

1. Capt. Drustrup quoted in "Mine Flood Is Halted: But Hope Dims For 12 Workers Still Entombed," *Sunday Independent*, January 25, 1959, sec. 1, p. 1, 2; Drustrup was also quoted in Joe Cooper, "January 22, 1959, A Day They Won't Forget," *Sunday Independent*, January 22, 1983, sec. 1, p. 14.

2. Frank Danna, taped interview, May 18, 1994, WVOHP, tape 1, side 1.

3. Arnold Embleton, taped interview, October 7, 1996, WVOHP, tape 1, side 1.

4. The exact number of gondolas used in the operation is less than clear. One newspaper story mentioned thirty-three, another thirty-eight, and yet another "about 55." *Newsweek* and *Coal Age* magazines said it was fifty ("She's Flooding," *Newsweek*, February 2, 1959, 18-19; "Pa. Mine Flood," *Coal Age* 64 (March 1959): 3, as did George Gushanas, the Glen Alden Coal Company superintendent who directed the riverfront effort to plug the hole. A caption to a photograph in the *Times Leader, The Evening News* reported that sixty gondolas were used ("More Coal Cars Dumped into Knox Mine," January 24, 1959: 3). One mine inspector said that thirty-eight were dropped into the abyss, and during one of the post-disaster hearings Knox Superintendent Groves cited the same number (Robert Groves testimony before the Joint Legislative Committee to Investigate the Knox Mine Disaster, March 12, 1959: 464). In the two locally written accounts of the disaster, Sporher (1984: 125) put the number at thirty while Roberts (1984: 146) said it was "a few."

To complicate matters, the Lehigh Valley Railroad billed the Commonwealth of Pennsylvania $20,520 for twenty gondolas, and when the state refused to pay, the railroad requested payment from the Pennsylvania Coal Company, whose directors agreed to reimburse the full amount. (The number of gondolas and their cost were discussed at the meeting of the board of directors of the Pennsylvania Coal Company on August 27, 1959, PCC Papers, Stockholders and Executive Committee Minutes, PHMC, MG 282, Minutes 1838-1971, February 20, 1953 - December 15, 1961, document no. 1021: 266-268.)

In the authors' judgment the correct number fell somewhere between fifty-five and sixty, although probably no more than forty found their way into the hole. On January 22, 1959, between twenty-five and thirty were pushed off the Lehigh Valley Railroad tracks, but no more than twenty of these came from the railroad itself. Up to thirty more were gathered from other companies over the next two days, an undetermined number dropped in to the hole by crane while several others were used as part of the fill along the river bank. At least two of the hoppers drifted far from the breach and sat downstream along the shore. (A former mine worker wanted to claim these two gondolas for scrap. See Joe Kutchka to Congressman Daniel J. Flood, David Lawrence Papers, PHMC, MG 191, General File, Department of Mines and Mineral Industries, Knox Mine Disaster, carton 2, General Correspondence, 1963-66.)

About 550 smaller mine cars were also used to stop the water flow, most dumped into the breach but some used to support the fill along the river bank.

5. George Gushanas, taped interview, January 13, 1994, WVOHP, tape 1, side 2. Gushanas said that the thirty by thirty-foot steel "mat" mentioned in the next paragraph was made by crews from the Glen Alden. It "cracked like tooth picks" when tossed into the hole due to the force of the water (telephone conversation with George Gushanas, March 11, 1997).

6. Tom Heffernan Jr., "Men, Machines Cut Rush Of River Into Knox Mine," *Sunday Independent*, January 25, 1959, sec. 1, p. 2.

7. Col. Stanley T. B. Johnson, chief of the Baltimore District Office, Army Corps of Engineers, cited the water volume figure in, "Pumping Started; Water Goes Back Into Susquehanna River," *Wilkes-Barre Record*, January 31, 1959, 1, 16. Maurice K. Goddard, secretary of Pennsylvania's Department of Forest and Waters, presented the one-sixth figure in, "Governor Asks Legislature To Provide Help: U.S. Mine Drainage Act Amendment To Be Sought," *Wilkes-Barre Record*, January 26, 1959, 1.

8. Henry A. Dierks, Walter L. Eaton, and Ralph H. Whaite, *Anthracite Mine-Flood Disaster: Breakthrough of Susquehanna River Into River Slope Mine, Knox Coal Company, Port Griffith, Pennsylvania*, (Washington: U.S. Bureau of Mines, 1960), 21.

9. Henry A. Dierks, Walter L. Eaton, and Ralph H. Whaite, *Anthracite Mine-Flood Disaster*, 1960, 9.

10. The water did not reach the Lackawanna Basin because that field's mines constituted a separate "bowl" of anthracite demarcated by the

aforementioned "Moosic Saddle."

11. Telegram quoted in, "Rescue Hope Fading for 12 Missing Men," *Times Leader, The Evening News*, January 24, 1959, 1.

12. Oscar J. Harvey and Harrison G. Smith, *A History of Wilkes-Barre*, vol. 5, (Wilkes-Barre: Smith Bennett Corporation, 1930), 68-69. These authors cited New York City's daily consumption as 846,900,000 gallons.

13. "Anthracite Water Problem Needs Drastic Action," *United Mine Workers Journal*, 65 (March 15, 1954), 5. The 1920s figure of 10.5 tons of water per ton of coal was cited by Harvey and Smith, *A History of Wilkes-Barre*, vol. 5, 69.

14. Thomas M. Beaney, Willard G. Ward, and John D. Edwards, *"Resume Relative to the Impact Upon the Economy of the Pittston and Wilkes-Barre Areas Caused by the Mine Flood Which Occurred at Knox Coal Company, January 22, 1959,"* Pennsylvania Department of Mines, February 19, 1959, 1-2, found in Beaney, Ward, and Edwards to Joseph T. Kennedy, Joseph T. Kennedy Papers, PHMC, MG 191, carton 2.

15. *Webster's Second International Dictionary* defines a cofferdam as, "A water tight enclosure . . . from which the water is pumped to expose the bottom (of a river, etc.)."

16. Frank Handley, taped interview, December 10, 1988, WVOHP, tape 1, side 1. On the first inspections of the mines see, "Muck-Filled Corridors Of Knox Mine Explored: No Trace Of Bodies," *Scranton Times*, April 2, 1959, 3, 8.

17. Frank Danna, taped interview, May 18, 1994. WVOHP, tape 1, side 1; see also Frank Danna, *The Life and Times of Frank P. Danna*, 1997, published by the author.

18. Henry A. Dierks, Walter L. Eaton, and Ralph H. Whaite, *Anthracite Mine-Flood Disaster*, 1960, 40. A detailed account of the sealing from the point of view of a key participant can be found in Frank Handley, taped interview, December 10, 1988, WVOHP, tape 1, side 1 and 2.

19. William Rachunis and Gerald W. Fortney, *Report of Major Mine Inundation Disaster, River Slope Mine*, 1959, 12.

20. Frank Handley, taped interview, December 10, 1988, WVOHP, tape 1, side 1. Emphasis is Handley's.

21. The federal government approved $2.45 million on March 18, 1959 through the Office of Civil Defense Mobilization. Because of limi-

64 tations regarding spending, including a one-year sunset clause, less than half the total allocation was actually spent. The state government presented bills totaling $1.5 million to the Knox Coal Company and the Pennsylvania Coal Company for costs incurred in the struggle to halt the flow of water into the River Slope. The Knox company paid nothing because it went bankrupt. The Pennsylvania Coal Company never paid its share of the bill, successfully presenting the legal argument that it was not liable. See "State To Present Bill For Knox Disaster Cost To Two Coal Companies," *Sunday Independent*, July 12, 1959, sec. 2, p. 1; "Justice Department Pressing Claim To Collect Knox Disaster Cost," *Sunday Independent*, September 25, 1960, sec. 1, p. 9.

Chapter Three
What Does it Mean?
How Could it Have Happened?

Dorrance Mine is Ordered Closed As Threat to Industry Increases: Economy of Entire County is Periled As Water Seeps from One Colliery to Another
   Sunday Independent headline, February 22, 1959[1]

[The] cause of the inrush of water was the removal of the natural support (coal) in the immediate vein beneath the river where the rock strata was insufficient to support the river. The contributory cause was the swollen ice-laden river.
   U.S. Bureau of Mines[2]

Immediate and Short-Term Consequences

Within twenty-four hours of the rupture, water inundated several mines adjacent to the River Slope, forcing the U.S. Bureau of Mines to close eleven operations. The Lehigh Valley Coal Company, the largest affected business, shut down the Henry Colliery and idled 650 workers.

The other ten closings involved small independents who held leases mostly from the Pennsylvania Coal Company. Together the closures affected over 1700 employees who had produced nearly 1.1 million tons of coal and earned $8.67 million in 1958. The Bureau of Mines requested that fifteen non-adjacent companies suspend work pending further investigation of the damage.

Heightening the immediate alarm was the knowledge among operators, inspectors, miners, and even ordinary citizens that the barriers—the one hundred-foot thick walls of solid coal separating one mine from another—were dangerously thin in certain places. As stipulated by the Pennsylvania Coal Mining Law of 1891, companies were required to leave the barriers to prevent a flood or other accident from spreading.[3] However, many of the subterranean palisades had been seriously weakened or robbed by the independent lessees as well as by the large firms. The barriers offered thousands of tons of "easy" coal to unscrupulous companies who were willing to violate the law in order to realize large profits. A missing barrier meant that water could easily travel from one working to another. A thinned wall, on the other hand, meant that water might build to very high levels and then burst into an adjacent mine with great force.

No one knew the extent of the piracy but it was fairly common knowledge that dozens of companies had engaged in the pillaging over decades of often loose and corrupt mining practices.[4] Not that barrier robbing was illegal in all cases. Sometime inspectors allowed it in higher-lying veins. For example, in 1930, the Conlon Coal Company, which mined under a lease from the Pennsylvania Coal Company, secured permission to break through the barrier separating the Pennsylvania's No. 14 Colliery from the Lehigh Valley's Enterprise Colliery. After the Knox disaster, however, as water levels climbed, this gap in the underground levee system endangered the Enterprise and the Henry mines.[5] But in most cases, the barriers were illegally taken. As one post-Knox disaster report stated:

> Some of these dangers should have been made impossible by barrier pillars. However, there is a great lack of faith in these barriers among mining men because of careless and greedy mining and because of the manner in which some mines have been leased out to completely irresponsible parties.[6]

Surprisingly, on Monday, January 26, 1959, the work schedule pub-

Figs. 33 Newspaper cartoon critical of anthracite mining practices. (Courtesy of *Sunday Independent*)

lished in local newspapers indicated that the Knox disaster had little effect on the region's main operators. The Glen Alden Corporation, anthracite's largest producer with many of its mines at the southern part of the field, worked as usual. All collieries at Susquehanna Collieries, producing mainly to the south, remained open. In the middle of the district, the Lehigh Valley continued operations at the non-adjacent Dorrance Colliery. The Hudson Coal Company also worked its few remaining mines. All non-adjacent independent companies scheduled work.

Nevertheless, operators and government officials faced a pressing question: how far would the subterranean flood range beyond the vicinity of the River Slope? By late February, the U.S. Geological Survey con-

Figs. 34. Newspaper cartoon critical of anthracite mining practices. (Courtesy of *Sunday Independent*)

cluded that pumping had lowered the water pool in the River Slope from 500 feet to 420 feet. Yet a large volume of "make water" still entered the mines and required continuous pumping. The Geological Survey also reported that the water level at an adjoining mine declined about seventy-eight feet from its highest reading, but it was impossible to determine how much of the drop resulted from pumping and how much from leakage across the barriers into other workings.[7]

Governor David Lawrence exacerbated public anxiety when he spoke ominously about the possibility of a second River Slope puncture, "which could produce far more devastating results than those experienced to date."[8] According to the governor, the river breach had not only washed away virtually all timber roof props in the River Slope Mine, but the churning of subsurface tides would continue to thin the ceilings in other chambers.

Obviously a regional underground flood would have devastating consequences for the hard coal industry. In spite of a decline in demand that had begun in the 1920s, anthracite still commanded a large share of the area's economy. The 1958 production of 7,669,440 tons had a market value of $76.9 million. Transportation costs of $4 per ton yielded another $30.7 million. With an unemployment rate over 11 percent—nearly twice the state figure—and a median income about 80 percent of the state average, the northern field desperately needed each of its 11,636 coal-related jobs.

A weakened industry would also present significant budgetary problems for local governments because coal companies still paid a large share of local taxes. In one worst-case projection, Luzerne County officials feared losing as much as 75 percent of the industry's $30 million tax assessment.[9]

Ground subsidence, or "caving," presented another post-disaster problem. Cave-ins had long been a regional problem in part because of two conditions in the northern field. First, the area stands as the most densely populated urbanized mining region in the United States. Typically, mining takes place in rural or relatively small isolated communities. But because of the northern field's geography and geology, an urban corridor developed along the Wyoming and Lackawanna valleys containing over one quarter million people in dozens of municipalities. With such a densely settled population, even a minor subsidence could cause extensive destruction.[10]

Second, the region's history of caving grew out of mining laws that separated surface rights from mineral rights and exempted the owner of the latter from liability for surface damage. Moreover, late in the nineteenth and early in the twentieth centuries, as the higher-lying veins became exhausted, the companies encouraged "second" and even "third" minings whereby they or their contractors removed, or robbed, the underground support pillars of solid coal which separated the numerous mine chambers. Indeed, the deadly Twin Shaft Disaster of 1896, where fifty-eight men were entombed at Pittston, resulted in large part from removal of support pillars. Pillar robbing radically altered the infrastructure of the region's mines. With the pillars gone, the mine roofs were weakened and surface areas began to collapse. Therefore, it came as little surprise when, within one week of the Knox rupture, a rash of subsidences struck Port Griffith. In the weeks following, cave-ins spread to communities up and down the Wyoming Valley. [11]

For the people of the valley, the Knox cataclysm caused great anxiety and created numerous immediate problems. However, as serious as the impending issues were, one question loomed just as large: how could it have happened?

How Did the Knox Mine Disaster Happen?

Strong emotional cross-currents gripped the community in the days and weeks following the disaster. Empathy for the victims and their families constituted one wave. The Port Griffith Disaster Fund Committee collected more than $12,000 from local charitable organizations, unions such as the International Ladies' Garment Workers (ILGWU) led by Min L. Matheson, and individual donors from throughout the coal fields and as far away as Canada, California, and Mexico. Several gifts came from a February 7th appeal by entertainer Perry Como during his nationally televised program. Como made the appeal at the request of Governor Lawrence. [12]

Another wave was characterized by tremendous anger and indignation. Its source went beyond the Knox catastrophe to a deeply-held sense of grievance and animosity toward the "coal barons" and what the anthracite years had done to the region. Here was another example of pain and sorrow caused by an exploitative industry and its greedy masters. Incensed journalists, politicians, clergy, and ordinary citizens demanded to know how such a mishap could have happened.

Although deep mining under a waterway was not prohibited by law, the practice nevertheless posed several dangers. The key engineering consideration required the maintenance of sufficient rock cover or roof within the mine. Regulations for mining under a water course required a roof of at least thirty-five feet and, ideally, fifty feet. Mining engineers determined the thickness of ceiling rock by drilling boreholes from the surface. Regulations also required companies to draw "stop lines" on mine maps, in red pencil, to indicate the boundaries beyond which mining could not take place because of inadequate rock protection.

Fig. 35. School damage in Port Griffith, Pennsylvania following the Knox Mine Disaster. (Courtesy of the Wyoming Historical and Geological Society.

The Knox Coal Company at first used the fifty-foot limit in its mines, following the policy of its lessor, the Pennsylvania Coal Company. However, Knox later wanted to use the lower figure of thirty-five feet at the River Slope because it allowed mining in formerly off-limit areas. The Pennsylvania company concurred with the change and redrew the stop lines to allow its lessee to mine further under the Susquehanna. Yet the highest ranking officials at Knox ignored even the redrawn stop lines and continued mining under the river until the roof shrank to a fatally thin margin somewhere between nineteen inches and a few feet. It was a gamble destined for infamy.

To complicate matters, two unique geological features of the Wyoming Basin made quarrying under the Susquehanna all the more difficult—the "Buried Valley" and the "Wyoming Anticline."

The Buried Valley of the Susquehanna and the Wyoming Basin Anticline

Between the river bottom and the ceiling rock of the anthracite coal beds lies a thick, heavy layer of sand, clay, and gravel that had been left by the erosive action of the Wisconsin ice sheet during the last ice age.[13] The deposits have been called "The Buried Valley of the Susquehanna." They extend to a depth of up to 320 feet, and follow the river for fifteen miles through the Wyoming Valley. These water-bearing, quicksand-like sediments have always presented a potential hazard for mining. Companies had to pay special attention to the roof cover separating the coal from the undulating hidden valley. Should the roof get too thin, the massive weight of the deposits could break through, causing a catastrophic inundation. The potential problem had been well recognized. In 1950 geologist S. H. Ash prophetically reported:

> If an opening should be driven from mine workings beneath the buried valley into the water-bearing deposits or if, because of subsidence, a cave[-in] should occur and water from the Susquehanna River flow suddenly into the mine workings, a major catastrophe could result. It is probable that a stream the size of the Susquehanna would resist all efforts to contain it in time to avert a large loss of life and could result in the loss of a major portion of the Northern field.[14]

Ash documented seventeen hazards in the Wyoming Valley area between 1872 and 1947 resulting from a failure in the rock strata underlying the buried valley. The most deadly, already mentioned in chapter 1, occurred on December 18, 1885, when twenty-six perished at Susquehanna Colliery Company's No. 1 Slope in Nanticoke. The only other mishap involving a flood-related death occurred in Wanamie in 1874 at the field's southern end.[15]

Fifty feet of ceiling rock emerged as the standard from a 1934 study by the state-appointed Water Hazards Commission. Composed of three mine inspectors, the commission was formed after a mine inspector found an independent lessee company working without adequate roof cover under the Lackawanna River in Scranton. Although the upper part of that river did not have the constraints of a buried valley, the lower part did. Moreover, numerous other independent coal companies had secured

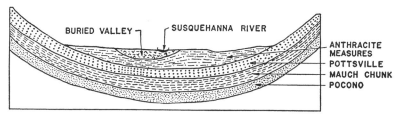

BURIED VALLEY
SUSQUEHANNA RIVER
ANTHRACITE MEASURES
POTTSVILLE
MAUCH CHUNK
POCONO
Figure 2.—Sketch section across coal basin, showing general relations of soft rocks of anthracite measures to hard, underlying Pottsville-Pocono rocks (after Darton).

Fig. 36. Cross section of the Buried Valley of the Susquehanna. (From W. Rachunis and G. W. Fortney, *Report of Major Mine Inundation Disaster, River Slope Mine,* U.S. Bureau of Mines, Wilkes-Barre, Pa., July 1959.)

similar leases and if even one of them mined with insufficient rock cover, a major inundation could occur. Acting under the authority of the 1891 state mining law, which required the Department of Mines to protect surface property as well as the health and safety of persons employed in and around the mines, the secretary of Mines appointed the commission to establish standards for mining under waterways.[16]

The commission worked closely with the engineering departments of ten coal companies. Its final report called for the creation of a safety zone in each mine where quarrying was prohibited. The report also reinforced the state mining law requirement that companies leave barriers:

> The Mining Laws are violated when an operator removes or weakens a barrier pillar legally established for the purpose of preventing the flow of water from one mine to another. The law is also violated when an operator carries on mining operations dangerously close to a body of water, either in a mine or on the surface.[17]

The commissioners recommended a mining prohibition where roof cover fell below fifty feet until boreholes had been drilled to ensure at least thirty-five feet. Although the recommendations were never embodied into law, they were adopted and enforced by the Department of Mines and applied to mining near all rivers and streams in the anthracite region.[18]

Because of these strictures, for some forty years before the Knox incident, all mining maps were marked with red stop lines beyond which mining was forbidden. Before leasing the River Slope property, the Pennsylvania Coal Company drilled approximately 100 boreholes and produced maps with the clearly indicated roof covers and boundary lines.

The Wyoming Basin's natural anticline presented the second geological confinement. Certain seams slant sharply toward the surface, topping out at a point dangerously close to the hidden valley, and then angle directly downward to create a "saddle." The effect is called an anticline. The breach at the River Slope occurred in a connecting chamber located precisely at the top of an anticline. Knox miners quarried the chamber in the Pittston vein to connect two larger tunnels that had been driven far off-course. The miners followed the steeply climbing chamber toward the river bed, to a point where the ceiling had only, at best, six to eight feet of rock cover topped by seventy-three feet of buried valley.

Illegal Mining Under the Susquehanna River

Because it was not a major company in control of mineral rights, Knox had to lease mines. Most of the company's previous leases permitted second minings to take pillars left behind from earlier excavations. With the 1954 River Slope lease, however, the company had the opportunity to extract virgin coal in one of the field's premier veins. The new mine reached high production levels. The last federal inspection before the break-in found an average daily output of 710 tons. In three daily shifts the company employed 174 men—151 underground and 23 on the surface.[19] The only problem was the mine's proximity to the river.

By September 1956 Knox wanted to take more of the Pittston Vein because the coal in the first lease was exhausted. In accordance with its lease agreement, President and General Manager Dougherty and Superintendent Groves directed William Receski, an assistant foreman, to prepare a "forecast" or a request for a permit to mine in a new area. Knox proposed to extend the River Slope by going 340 feet deeper into the ground at the same angle. The company would then dig a 124-foot chamber branching off the extension into solid coal. Most of the new mining would occur under the river.

Pennsylvania Coal Company authorities Ralph Fries and Fritz Renner examined the forecast and approved it. However, they had a problem. The new mining would take the lessee into forbidden territory beyond the stop line that had appeared on mine maps for the previous four decades. In a transparent effort to approve the mining *and* keep the slope extension "legal," they redrew the stop line and stretched it well out into the Susquehanna.[20] The redrawn boundary, though, presented Fries and Renner with yet another problem. Because the proposed mining area had not been tested with boreholes, the rock cover became uncertain and unknown. The two nearest boreholes showed large differences in covering—one read forty-nine feet, the other nineteen inches. In later testimony before state investigators, Fries and Renner argued that they used "interpolation" (an educated guess) to estimate a rock thickness of nearly thirty-five feet in the area between the two boreholes where most of the mining took place.

Rockmen completed the extension in about eighteen months. During the latter part of May 1958, Receski supervised the work on the new chamber which branched from the extension. However, the chamber was not quarried according to plan. It traveled much further than requested—not 124 feet but 260 feet, all of it in solid coal over eleven feet high. Just as importantly, Receski and his men diverted the tunnel twenty-three degrees to the west so that it did not run parallel to the River Slope but, in clear violation of the forecast, coursed far out under the Susquehanna, 125 feet of it lying beyond even the new stop line. (see fig. 37)

On August 28, 1958, Knox requested, and the same Pennsylvania Coal Company officials granted, permission to excavate a second 125-foot chamber parallel to the first. Still no boreholes were dug. To further complicate matters, company mine maps were not updated in a timely fashion so, according to Pennsylvania officials, they did not know the first chamber was improperly laid. With Receski as the supervisor, miners dug the second chamber much further than requested—not 125 feet but about 260 feet—and it traveled 160 feet beyond the revised stop line. Moreover, because of the steep anticline in this part of the field, this chamber climbed seventy-three feet over a horizontal distance of 160 feet at a thirty-degree angle. It came within seventy-five feet of the borehole showing nineteen inches of rock cover topped by thirteen feet of buried valley. The two places were completed in late July or early August 1958.

After excavating the two lengthy chambers, Receski said that he received "verbal permission" from Fries to join them with three smaller interconnecting or crosscut chambers. In testimony before investigative bodies, Fries denied that he had given the verbal permission. That Knox

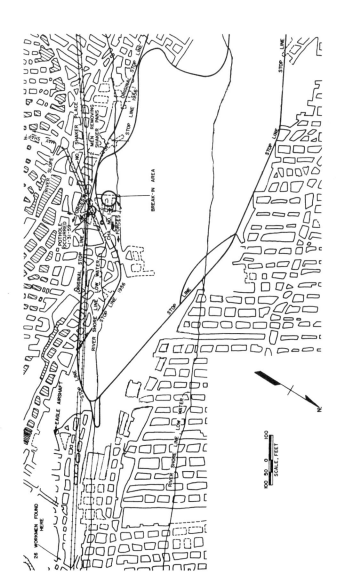

Fig. 37. The River Slope Mine showing stop lines. (From W. Rachunis and G. W. Fortney, *Report of Major Mine Inundation Disaster, River Slope Mine*, U.S. Bureau of Mines, Wilkes-Barre, Pa., July 1959)

Fig. 38. General offices of the Pennsylvania Coal Company, Dunmore, Pa. (Courtesy of National Canal Museum, Hugh Moore Park)

did not prepare a forecast requesting approval for the crosscuts indicated the suspicious nature of the mining. To get the maximum output, Receski had his miners widen one of the crosscuts to twenty-seven feet instead of the usual twelve feet. Such broadness provided an even greater risk for roof failure. Crucially, one of the crosscuts traveled up the anticline to a point where the rock cover between the chamber and the buried valley stood at no more than six to eight feet. The river broke into this cavity.[21]

State and federal mine inspectors apparently missed the illegal mining until January 13th when they ordered the mine shut down. Pennsylvania company inspector Joe Stella had discovered the situation much earlier and reported it to his superiors; but they took no restrictive action. Knox officials most likely knew that they had ventured into illegal territory, but the allure of the rich vein evidently proved irresistible.

Consequently, the immediate cause of the Knox catastrophe could be attributed to off-course mining under the Susquehanna River past the original and even the revised stop line, specifically in a crosscut that followed the anticline to a perilously high level. The entire area was mined without the benefit of boreholes or accurate mine maps. The added weight of the rising Susquehanna on January 22, coupled with a ceiling of very thin rock, caused the buried valley and the river to collapse into the mine.

During the first day of the tragedy, observers described it as a true accident, "another in the succession of jolts which Pennsylvania withstood throughout the day, all the result of freakish weather in January [where] creeks and rivers boiled over their banks [and] ice broke up in many different streams, all causing extensive damage."[22] However, in the following weeks that view changed and serious questions arose about blame and responsibility. Rumors of negligence, greed, and even company-union corruption emerged. The public began demanding answers. The son of one of the missing miners wrote a letter to the editor (name withheld by request) that captured much of the citizenry's frustration and anger:

> I have paced the floor to find words to start this letter, but it was very hard knowing my dad is trapped in the Knox Coal Co. mines. We still have faith and hope in God that they will come out alive. I will not rest until I know all the answers to a few questions I have to ask.
>
> 1. Is it a mining law that mines should have escape outlet? If so where was the escape outlet at Knox Coal Co.? If it was the Eagle shaft, why was it covered with debris?
>
> 2. Did the mine have any warning signal?
>
> 3. How come the tunnel was only 30 feet under the [bed of the] Susquehanna River? Was this the safety limit? What kind of rock strata is under the river and how far down do you have to be sure that it is safe for mining?
>
> 4. What is the exact complete ownership-managerial setup of the Knox Coal Co.? Is someone hiding something?
>
> 5. This is the last question I have at the moment. How come the long delay in pumping operations?
>
> The only way to avert any more mine disasters of this magnitude is for the working men of Wyoming Valley to band together and demand the truth. Be men and don't be afraid to speak the truth.[23]

Notes to Chapter Three

1. Headline, *Sunday Independent,* February 22, 1959, sec. 1, p 1, 2.

2. William Rachunis and Gerald W. Fortney, *Report of Major Mine Inundation Disaster, River Slope Mine, May Shaft Section, Schooley Colliery, Knox Coal Company, Incorporated, Port Griffith, Luzerne County, Pennsylvania* (Wilkes-Barre, PA: U.S. Bureau of Mines, 1959), 15.

3. *Coal Mining Laws of Pennsylvania* (Washington, DC: USGPO, 1891), 12. The exact thickness of the barrier was not stipulated in the original law, but emerged later.

4. State authorities allowed companies to dig interconnecting passageways and conduct some "development work" in higher lying sections of the barriers. For this reason, miners have spoken about the possibility of walking underground from one end of the northern field to the other.

Tony Waitkevich discussed how he participated in robbing the barrier for the Knox Company in his taped interview, August 2, 1989, WVOHP, side 1, tape 1. Several other mine workers discussed the barrier robbing in oral history interviews. See, for example, William Hastie's discussion of the practice in his taped interview, July 31, 1989, WVOHP, tape 1, side 1. Along with lessee firms after the 1930s, so-called bootleg miners who mined illegally during the 1930s were known to rob barriers. On bootleg mining see Pennsylvania Anthracite Coal Industry Commission, *Bootlegging or Illegal Mining of Anthracite Coal in Pennsylvania, A Census and Survey of the Facts* (Harrisburg, 1937).

Bootleg mining was more prevalent in the southern and middle coal fields, although it also occurred in the northern field. On bootleg mining in the Moosic area see Charles Orlowski, taped interview, WVOHP, June 9, 1998, tape 1, side 1.

5. "Lehigh Valley Plans To Close Dorrance Mine Again This Week," *Times Leader*, February 23, 1959, 3.

6. "Dorrance Mine Is Ordered Closed as Threat to Industry Increases," *Sunday Independent,* February 22, 1959, sec. 1, p. 1, 2. Another news report discussed the weakened barriers: "Adjoining the flooded Henry-Prospect mines, which have been flooded and are the cause of the Dorrance closing, is the Old Miners Mills Coal Co., the door through which flood waters can empty and fill one mine after another down the line to and including the Bliss and across to the West Side of the Susquehanna River.

There is no strong barrier pillar between Henry-Prospect and Old Miners Mills Co. Following down the line in succession, no heavy barrier is left to protect Pine Ridge, Mineral Springs, Baltimore 5, Hollenback, South Wilkes-Barre, Huber, Sugar Notch, Truesdale and Bliss Collieries. The underground stream, unless checked by the many pumps operating in these mines, could continue all the way to Wanamie, which has a barrier pillar 200 feet thick. Mining engineers point out that because of the connecting links of collieries and absence of barrier pillars, the Knox Coal Company waters eventually can reach into Loree and Lance Collieries in Larksville, [and the] Nottingham and Avondale in Plymouth." (See "Here's How Flooding Through Chain Reaction Can Ruin Mines in Area," *Sunday Independent,* February 22, 1959, sec. 1, p. 2.)

7. "Fears More Mine Floods: Official of LV is Pessimistic," *Scranton Tribune,* February 25, 1959, 3, 13.

8. "Water Threat in Coal Mine Still Present," *Scranton Tribune,* February 25, 1959, 3-13.

9. "$30 Million Loss in Assessments Feared," *Times Leader, The Evening News,* February 27, 1959, 3. A tax assessor challenged this figure as being too high. See "Clark Terms Wood's View 'Ridiculous': $30 Million Loss in Valuation Due to Mine Flooding Seen 'Absurd'," *Times Leader, The Evening News,* February 27, 1959, 3.

10. According to U.S. Census figures Luzerne County's 1960 population count stood at 346,972 people living in seventy-four municipalities, while the comparable figures for Lackawanna County were 234,331 and forty respectively. Not all of the municipalities were underlain with coal, but the most populous ones were.

11. For a thorough discussion of the northern field's subsidence history before 1948 see Ellis W. Roberts, "A History of Land Subsidence and Its Consequences Caused by the Mining of Anthracite Coal in Luzerne County, Pennsylvania," unpublished doctoral dissertation, School of Education, New York University, 1948.

The northern region used the "room-and-pillar method" of mining whereby relatively large "rooms" of coal were removed but support pillars of coal up to thirty feet thick were left standing between them. A second or a third mining involved removal of the support pillars (see Appendix II).

12. On the garment workers' assistance see "ILG Rushes Pittston

Flood Aid," *Justice* (International Ladies' Garment Workers' Union magazine), February 1, 1959, 1. The spouses of two Knox victims, Mrs. Altieri and Mrs. Kaveliskie, worked in garment manufacturing, as did John Baloga's son. Ten of the rescued miners' wives and daughters were employed in the garment industry. On Perry Como's appeal see "30 Years Ago: Investigators Looking Into Knox Mine Disaster," *Sunday Dispatch*, January 22, 1989, 2.

As mining employment declined during the 1940s and 1950s the garment industry became a valuable source of family income in the anthracite region, particularly for women. For an account of garment union organizing during this period, see Robert P. Wolensky and Kenneth C. Wolensky, "Min Matheson and the ILGWU in the Northern Anthracite Region," *Pennsylvania History: Special Issue on Oral History* 60 (1993): 455 474; Kenneth C. Wolensky and Robert P. Wolensky, "Building the ILGWU in Pennsylvania's Anthracite Mining Towns: The Leadership of Min Matheson, 1944-1963," *Sociological Imagination* 31 (1994): 83-100.

13. S. H. Ash, "Buried Valley of the Susquehanna River, Anthracite Region of Pennsylvania," (U.S. Bureau of Mines, Bulletin 494, Washington, DC: USGPO, 1950), 7. See also S. H. Ash, W. L. Eaton, and others, *Flood Prevention Projects at Pennsylvania Anthracite Mines*. Progress Report for Fiscal Year Ending June 30. 1947 (USBM: Report of Investigations 4288, 1948); S.H. Ash, S.B. Davies, and Others, "Barrier Pillars in Lackawanna Basin, Northern Field, Anthracite Region of Pennsylvania (USBM: Bulletin 517, Washington, DC, 1952).

14. S. H. Ash, "Buried Valley of the Susquehanna River, Anthracite Region of Pennsylvania," 1950, 2.

15. The buried valley also extends a short distance into the Lackawanna River which joins the Susquehanna above Pittston. Ash reported two water-related accidents in the Lackawanna area, one in 1884 and another in 1885, neither resulting in death. The disasters are listed in Ash, "Buried Valley of the Susquehanna River, Anthracite Region of Pennsylvania," 1950, table 1, p. 13. Ash, Davies, and others in "Barrier Pillars in Lackawanna Basin, Northern Field, Anthracite Region of Pennsylvania," 1952, estimated that about one-half of the reserves in Lackawanna County had been removed from production because of underground flooding.

16. The Water Hazards Commission's final report was presented in S. J. Phillips (State Mine Inspector and Secretary of the Water Hazards

Commission), "Water Hazards in the Lackawanna Valley," Anthracite Section of the American Institute of Mining and Metallurgical Engineers, June 29, 1935.

17. S. J. Phillips, "Water Hazards in the Lackawanna Valley," 7.

18. S. J. Phillips, "Water Hazards in the Lackawanna Valley," 23.

19. William Rachunis and Gerald W. Fortney, *Report of Major Mine Inundation Disaster, River Slope Mine*, 1959, 5.

20. Thomas M. Beaney, Willard G. Ward, and John D. Edwards, *Report of the Pennsylvania Commission of Mine Inspectors* (Harrisburg: Pennsylvania Department of Mines and Mineral Industries, April 7, 1959), 13.

21. *Report of the Commission of State Mine Inspectors* cited the six to eight foot roof thickness which was discovered by examining a similar but smaller surface break through at the River Slope after the cofferdam had been built (April 7, 1959), 3. This figure corrects the widely reported nineteen inches of cover, a figure that no doubt came from the thickness of rock at the nearby Borehole No. 1146. The *Report* also stated that the breach occurred seventy feet from the Slope (p. 9).

22. "12 Missing and 35 Rescued After River Breaks into Mine Near Pittston," *Wilkes-Barre Record*, January 23, 1959, 1, 2.

23. "Son of Missing Miner Wants Answers from Knox Owners," *Wilkes-Barre Record*, January 28, 1959, 13.

Chapter Four
Establishing Blame and Responsibility

That's all, just greed. You can name a book Greed and write forever on it in the Anthracite, from the first mines that were ever opened. That's all it was, all the time. Did anybody ever leave anything for a coal miner? Is there a park ever donated by a coal company? Not one. If it wasn't for Kirby Park—which was not coal connected really although it was railroad—[we'd have] nothing. They left nothing for you. All the streets are named after them. Half the streets in Port Griffith are named after former Pennsylvania Coal Company officials. All the shafts were named after them. And they never left anything.
Al Kanaar, Knox Coal Company miner,
Inkerman, Pennsylvania[1]

Somebody was certainly negligent somewhere
State Rep. Frank Kopriver, Allegheny County[2]

As details of the Knox disaster became more widely known, it became clear that "accident" hardly seemed an appropriate characterization. Victims' families, letter writers, editorialists, and public officials demanded answers to many pressing questions, including the key one: Who supervised such reckless mining?

The disaster opened the door to tough questions about mining legislation. Were weak or outdated laws responsible? Were the existing statutes adequately enforced? Scrutiny also turned to the mining industry: Was there something in the operation of this declining business that led to the misadventure? On this point one editorialist asked: "Why is it possible for small mining companies to operate profitably in mines that had been previously abandoned by larger companies as unprofitable?"[3] And where was the United Mine Workers of America, its safety committees, and grievance processes when these violations were occurring? Even more pointedly, was there any truth to the widely circulating rumor that Knox and UMWA officials had established a sweetheart contract that pushed safety aside?[4] Another editorialist captured the public mood:

> The time has come to find out what is going on behind the scenes in the hard coal business. Perhaps this disaster with its serious implications will provide the lever to pry off the lid.[5]

However, establishing culpability would be no easy task, for mining accidents had always been a difficult subject in the anthracite region. When 10,110 mineworkers died in mishaps between 1870 and 1900, another 14,998 between 1901 and 1930, and a grant total of 35,000 throughout anthracite's history, the companies and their officials remained virtually blameless. Instead, the workers were most often cited for having a careless disregard of safety.[6] Similarly, when subsidence caused extensive damage to surface structures in the Wyoming and Lackawanna valleys, the companies escaped liability thanks to legal interpretations by the courts.

Nevertheless, because the death and destruction in Port Griffith caused such an egregious breach not only of the Susquehanna River but of public confidence in the coal enterprise and in government's ability to regulate it, resolute demands for answers—and justice—could not be ignored. State and federal government authorities consequently undertook four investigations during the months following the calamity.

Fig. 39. Joint Committee hearings, St. Cecilia's Church, Exeter.
(Courtesy of *Scranton Times*)

FOUR INQUESTS INTO THE KNOX MINE DISASTER AND THE ANTHRACITE COAL INDUSTRY

The Pennsylvania Legislature established the Joint Committee to Investigate the Knox Mine Disaster, which initiated the largest inquiry. Composed of five members from both General Assembly chambers, the Joint Committee was charged with determining the causes of the catastrophe and proposing measures to prevent a similar future occurrence.[7] Sixty-four witnesses including owners, UMWA officials, miners, laborers, and inspectors were called to the committee's four hearings at St. Cecilia's Church in Exeter during February, March, and April 1959. Their testimony filled 1,617 pages of transcript.

The committee uncovered a trail of greed, risk-taking, and negligence. It sat in anger as owners Dougherty and Fabrizio, as well as union official August J. Lippi, invoked the Fifth Amendment even though they were not on trial. The owners' defiance so angered lawmakers that Fabrizio and Dougherty were ordered to testify in Harrisburg before the entire Senate—the first time in 202 years that the Pennsylvania legislative body had issued such an order to a citizen.[8]

Fig. 40. "August J. Lippi Takes Oath, But Then Clams Up." (Courtesy of *Sunday Independent*)

The committee questioned fearful employees who were reluctant to speak. It listened as state and federal inspectors argued that falsified maps and other deceptions prevented them from conducting accurate surveys. It heard the president of the Pennsylvania Coal Company, Joseph A. Martin, disavow any liability or responsibility for the "accident."

The committee's *Final Report* contained a somewhat surprising conclusion, however. Although it acknowledged wrongdoing on behalf of the Knox company, it placed greater emphasis on a seemingly impersonal "chain of events each link of which was necessary or incidental to the resultant tragedy."

> It is the opinion of this Committee that a combination of contributing factors including, but not limited to, indifference and apathy on the part of the lessor concerning the mode of extracting coal; incompetence on the part of the supervisory employees of the lessee; an incentive pay system resulting in the miners and other incentive employees conducting mining without sufficient regard for their own safety or position; owners of the lessee uninformed and uninter-

ested in the means of extracting coal; a lack of qualified professional engineers and/or consultants on the staffs of the lessor or the lessee."[9]

Two other links in the chain were the sudden rise of the river and "an archaic Anthracite Act." To address the latter, the committee recommended "a thorough and comprehensive re-evaluation and investigation of Pennsylvania's present Mining Laws. . . ."[10] However, critics of the report such as Luzerne County District Attorney Albert Aston did not easily accept the "chain" argument:

> The committee report, distilled to its essence, admits to a disaster, places no blame, and suggests remedial legislation. We mean no disrespect when we say it is almost a case of locking the barn door after the horse has been stolen. . . . The people of Luzerne County rightfully demand that where evidence points to criminal negligence, those responsible be compelled to make legal answer."[11]

Nonetheless, the information gathered by the body was used in efforts to reform state mining laws and in future criminal proceedings.[12]

The Commission of State Mine Inspectors conducted the second investigation.[13] This three-member panel consisted of veteran mine inspectors who held hearings on February 16-17, and March 24, 1959. They examined the causes of the break-in and looked for instances of criminal negligence. Twenty-two witnesses testified. Representatives from the federal Bureau of Mines participated in the inquest and used the hearings to write a separate report.

The commission made two recommendations: prosecute appropriate persons and enact legislation to forbid mining under a waterway without approval from the Department of Mines. Seven persons were accused of violating state mining laws: Robert Groves, superintendent; Frank Handley, foreman; William Receski, James Jeffrey, and Simon Zaboroski, assistant mine foremen; and Fritz Renner and Ralph Fries, engineers at the Pennsylvania Coal Company. No state or federal mine inspectors were implicated.

The U.S. Department of the Interior's Bureau of Mines supervised the third inquiry. The strength of this report lay in its technical analysis and graphic illustrations.[14] The bureau said nothing about illegal mining, indictments, or prosecutions. It endorsed a mining prohibition in all subterranean seams with insufficient rock cover; a mining prohibition beyond an established stop line; a mining prohibition without a written permit; closer spacing of boreholes; and keeping all mine surface openings clear because they may be needed as emergency outlets.

The fourth inquisition proved to be the most important and explosive. It actually began several months before the Knox plight when a federal grand jury convened in Scranton to examine organized crime's influence in northeastern Pennsylvania's industries including anthracite mining and garment manufacturing. The body worked with the Federal Bureau of Investigation, the Internal Revenue Service, and the recently created Special Group on Organized Crime within the U.S. Department of Justice.[15]

The U.S. attorney general created the Special Group following the historic raid on the November 14, 1957, gangland convention in Apalachin, New York. This meeting took place on an estate about seventy miles north of Scranton owned by Joseph Barbara whom authorities described as the former top boss of organized crime in northeastern Pennsylvania.[16] The Special Group had begun working with the grand jury because the anthracite mining business was apparently on the Apalachin agenda (as was the control of garment factories in Pittston), and because four of the fifty-eight reputed mobsters present resided in the Pittston area. One of them was Dominick J. Alaimo, an official at one of the UMWA's larger locals, No. 8005, which covered ten mining operations in Luzerne and Lackawanna Counties—including the Knox Coal Company's River Slope.[17]

On February 18, 1959, the grand jury began scrutinizing the anthracite industry, including but extending far beyond the Knox calamity. Chief Investigator Thomas J. Brennan of the Special Group's New York office hoped that the evidence would "open the lid on [criminal] mining industry activities, especially in the Pittston area."[18] Virtually all of the first group of witnesses had some affiliation with the Knox Coal Company or District 1 of the UMWA which encompassed all of the locals within the northern field.

After hearing from over thirty persons and reviewing voluminous records from the Knox and Pennsylvania Coal companies, District 1 UMWA, and three local banks, in March 1959 the grand jury handed down indictments against the Knox Coal Company and six individuals.

Robert Dougherty, president and general manager, received a twenty-five count indictment, and Louis Fabrizio, secretary-treasurer, took a nine-count indictment—both for Taft-Hartley labor law violations. They were specifically accused of bribing union official Dominick Alaimo a total of $1,800 and $7,000 respectively, to allow the company to violate the labor-management contract regarding pay, safety, and other work conditions. The Knox Coal Company was hit with a separate thirty-four count indictment for the same reason.

Alaimo, on the other hand, was charged with thirty-four counts of receiving the illegal payments.[19] Thirty-four count indictments for labor bribes were also thrown at two other Local 8005 officials. They were Charles Piasecki, president, and Anthony Argo, committeeman.[20]

Before the grand jury recessed in August 1959, it indicted fifty-eight year-old August J. Lippi, District 1 president, for accepting $10,117 in illegal bribe payments from Knox officials. The charges alleged corruption at the highest levels of a union once cherished by mineworkers as much as it was feared by operators.[21] With such high ranking officials "on the take," safety and grievance procedures would certainly not rank as high priorities at the Knox Coal Company.

Dougherty, Lippi and Fabrizio secured a change of venue to Wilmington, Delaware, for their trials. Fabrizio's trial ended in a hung jury. The government retried the case some months later and the new jury rendered a verdict of not guilty.[22] Dougherty's trial also ended in a hung jury. The case was not retried.

Lippi's trial for bribery took place in June 1960. His argument that the money he received represented the repayment of a loan he had made to John Sciandra (an original owner who died in 1950) did not convince the jury. He was convicted on all three counts. However, the judge granted an appeal.[23] The U.S. Circuit Court heard the appeal and acquitted Lippi of the charges.

The three local union officials did not request a change of venue. At their trial in Scranton, Alaimo and Piasecki were found guilty and handed two year prison sentences plus a fine. Argo pleaded guilty and received a suspended sentence. Alaimo, Piasecki, and Argo were subsequently barred from the UMWA.[24] Paradoxically, the union officials who took the money were found guilty, but the company officials who offered the money were found innocent.

Fig. 41. Anthony Argo and Charles Piasecki face federal prosecutors. (Courtesy of *Times Leader*)

OTHER INDICTMENTS AND TRIALS

In addition to the federal charges, the Commonwealth of Pennsylvania, under the direction of the newly elected Luzerne County District Attorney Stephen A. Teller, convened a grand jury. It handed down indictments against Lippi, Fabrizio, and Dougherty, as well as Superintendent Robert Groves, Assistant Foreman William Receski, and Fritz Renner and Ralph Fries of the Pennsylvania Coal Company. Lippi, Fabrizio, and Dougherty faced three counts of conspiracy related to mining and labor law violations. They were also each charged with twelve counts of involuntary manslaughter. Groves and Receski were indicted on twelve counts of involuntary manslaughter. Renner and Fries faced the involuntary manslaughter charges plus a citation for violating state mining laws.

Fabrizio, Dougherty, and Lippi as well as Renner and Fries secured a state Supreme Court-mandated change of venue from Wilkes-Barre to Easton in Northampton County. In July 1960, a jury found all three guilty of manslaughter. Lippi and Fabrizio were also found guilty of conspiracy, but Dougherty was deemed not guilty on this charge. Pennsylvania Coal Company officials Renner and Fries were found not guilty of involuntary manslaughter and mining law violations.[25] However, the presiding judge granted the guilty defendants' motions in arrest of judgment which effectively threw out the conviction. The Commonwealth appealed. A Pennsylvania Superior Court then reversed the conspiracy and manslaughter convictions of Lippi, Fabrizio, and Dougherty. Another Commonwealth petition to reinstate the conspiracy charges proved unsuccessful.[26]

Perhaps the most shocking and troublesome finding during the Commonwealth's inquest involved August J. Lippi's proprietary interest in the Knox Coal Company. In clear violation of the Taft-Hartley labor law, the trial revealed that he owned stock in the company even as his union represented its employees. District Attorney Stephen Teller made the discovery as he cross-examined George Daileda, the head cashier at the Exeter National Bank, where Lippi served as president. Teller recalled a crucial moment in the trial:

> I remember one outstanding moment when I think back about the trial. August Lippi had endeavored to hide his co-ownership in the Knox mine. I must say that I pray each day that I won't say an evil word against anyone, and what I'm about to say is all a matter of public record.
>
> He was president, a large owner of the First National Bank of Exeter....
>
> When the cashier got on the witness stand in the trial, I showed him a check dated a certain date for $4,000, and I said, "This bears your mark, that you cashed the check." He said, "Yes, I cashed that check." I said, "That check was $4,000 dollars payable to Dougherty, was it not?" "Yes." I showed him a second check payable to Fabrizio dated the same date

Fig. 42. Robert L. Dougherty, on trial in Wilmington, Delaware for violating the Taft-Hartley labor law. From left to right: attorney Thomas Burke, co-counsel; attorney Arthur Maguire, co-counsel; Dougherty; Robert L. Dougherty Jr.; attorney B. Todd Maguire, co-counsel. (Courtesy of *Scranton Times*)

with his mark on it. And I said, "You cashed this check for $4,000 for Fabrizio did you not, it's the same amount as the check to Dougherty?" And see, those checks were endorsed, one by Dougherty, and one by Fabrizio. Now, I said, "I'm going to show you a third check. This is made out to cash it bears no endorsement on it. But you cashed it. Your number is on the check." I said, "It bears the same date as the other two checks, the same amount, $4,000." I said, "Who cashed that check?"

He remained silent. He didn't say anything. And there was total silence. The judge remained silent. The jury was silent. The spectators were silent. The witnesses gathered there were silent. And I thought to myself sometimes silence is more effective than the loudest noise. So I waited and the seconds passed by. The minutes passed by. The judge waited. Finally I said, "Tell us, tell us please, who cashed that check."

And then in a faint voice he said, "August Lippi." That

Fig. 43. Louis Fabrizio testifies at state investigation. (Courtesy of *Scranton Times*)

and other evidence tied August Lippi as a co-owner of the Knox Mine.[27]

Meanwhile, at the Luzerne County Court House in Wilkes-Barre, Groves and Receski, who had not requested a change of venue, went on trial before Judge Bernard J. Brominski for involuntary manslaughter. The trial lasted several weeks. The jury's verdict: not guilty. The defendants argued that they were merely following orders from superiors. According to attorneys Teller, the prosecutor, and William Digillio, the defense counsel, the carefully planned strategy that sent five higher ranking officials to Easton while keeping the two lower ranking figures in Wilkes-Barre had worked for Groves and Receski, and for the other three on appeal. The underlings blamed the superiors, and the superiors blamed the underlings.[28] In the final analysis, no one was ever judged criminally negligent in the twelve deaths at the River Slope Mine.

Although three of the four Knox owners (the fourth owner was Mrs. Josephine Sciandra, heiress of her husband's stock in the company, who was not indicted in these early cases) escaped convictions on Taft-Hartley, manslaughter, and conspiracy charges, they were convicted on other charges. Fabrizio pleaded guilty in two separate trials dealing with personal and corporate tax evasion associated with his income from the Knox Coal Company. He was sentenced to 181 days in Danbury (Ct.) federal

Fig. 44. Luzerne County District Attorney Stephen A. Teller and his prosecutorial team. From left to right: Vincent Quinn, first assistant district attorney; Charles Casper, assistant district attorney; Teller. (Courtesy of *Times Leader*)

prison on January 24, 1964, where he served nearly five months.[29]

Dougherty, who at age twenty-five had become the youngest mine foreman in Pennsylvania in 1924, pleaded no contest of evading $75,542 in personal taxes in 1957. The conviction resulted from a trial in Scranton that consolidated three conspiracy and income tax evasion cases against the defendant. In handing the sixty-six-year-old businessman a sentence of one year and one day, plus a $10,000 fine, the judge placed "great reliance on your [ten] character witnesses that you led a good life and contributed your part to charitable and civic endeavors." Dougherty was released after serving four and one-half months in the federal prison at Danbury, part of it in the company of Fabrizio.[30] In a related case, two other coal companies owned by Dougherty, the Peeley and the Avon, were judged guilty of evading income taxes, although he was acquitted in that case.[31]

Josephine Sciandra was convicted on one count of conspiracy to avoid

Fig. 45. Robert Groves and William Receski and their attorneys at the Luzerne County Court House. From left to right: attorney Frank McGuigan, co-counsel for Groves; Receski; Groves; attorney William Degillio, counsel for Receski; attorney John Mulhall, co-counsel for Groves. (Courtesy of *Times Leader*)

corporate income taxes, one count of engaging in illegal acts to avoid income taxes, and two counts of personal income tax evasion. She was fined $2,000 and placed on three years' probation.

Additional Indictments Against August J. Lippi: Scandal in District 1

August J. Lippi was hit with several additional indictments which led to three other trials. Two of the cases were for corporate and personal income tax evasion, the other for bank fraud in his role as president of the Exeter National Bank. Lippi was found guilty in one income tax trial and not guilty in the other. He received a sentence of three years in prison plus three years of probation, and a $5,000 fine. The federal courts denied an appeal. Within hours of the denial, Lippi was arrested by the FBI

Fig. 46. Luzerne County Judge Bernard J. Brominski (1991 photo), presided over the Groves and Receski manslaughter trail. (Photo by Robert P. Wolensky)

at New York's Idlewild Airport as he attempted to leave the country aboard a flight bound for South America.[32]

While earlier convictions were on appeal, federal agents again arrested Lippi as he attended a union meeting in Washington. He was charged with thirty-four counts of conspiring to defraud the Exeter National Bank of $39,000. Although examiners actually discovered the misuse of over $400,000, only $39,000 was actually traceable to Lippi's account. The bank's head cashier, George Daileda, pleaded guilty to thirty-four counts of bank fraud and, despite threats to his safety and that of his family, turned government's witness against Lippi. During the course of the trial, Daileda's son, Norman, was shot in the leg by a high-powered rifle as he worked late into the evening at the family's Exeter insurance agency. The unknown assailant fired through the main door of the building. Law enforcement authorities, interpreting the attack as a threat and a retaliation against the senior Daileda for turning government witness, provided around-the-clock protection for the Daileda family. George Daileda continued with his testimony.[33] Lippi was convicted and sentenced to five years imprisonment plus five years of probation and a $10,000 fine. The U.S. Court of Appeals affirmed the conviction. The U.S. Supreme Court denied a review.[34]

Despite impassioned pleas regarding his diabetes and heart disease, as well as claims that he was capable of rehabilitation without jail, after a few years of appeals and other delays, the former district president was remanded to the Lewisburg federal prison on November 5, 1965. The five year sentence for bank fraud ran concurrently with the three-year sentence for income tax evasion. At least part of Lippi's incarceration at Lewisburg would be spent in the company of another fallen labor leader, James R. Hoffa, of the International Brotherhood of Teamsters who would later become a secret investor in another corrupt anthracite mining company, Blue Coal.

Even with a string of indictments, convictions, and appeals, Mr. Lippi's position as District 1 president remained secure. Following the Taft-Hartley bribery charges, UMWA President John L. Lewis, and Vice President, Thomas Kennedy, refused to comment until the union conducted its own review. The union's silence toward the Lippi affair angered the local press:

> This is getting hard to believe. Now we have seen what is

at least purported to be rather complete evidence that August Lippi, president of the local district of the United Mine Workers, was part of the Knox Coal Company ownership. This would seem to justify the suspicion that he was playing both the operators' and the union's side at the same time. Yet—and here's the unbelievable part—not the slightest word even of comment has come from the leadership of the UMW.[35]

In August 1960, Lippi received the nomination for another four year term, the protests by some rank-and-file members about one set of convictions (which were on appeal) notwithstanding. UMWA headquarters in Washington said it would only intercede in the election if the convictions were upheld.[36] Lippi was re-elected by a wide margin.

In March 1965, following one guilty verdict but prior to his sentencing, Lippi was re-nominated for yet another term as president at the District 1 convention in Scranton. Indeed, the 125 convention delegates voted to raise his salary from $17,000 to $20,000. The pay increase came without regard to the district's reliance on subsidies from the union's international office or its continuing membership decline.[37] Immediately prior to the convention, the new UMWA president, W.A. "Tony" Boyle, named Lippi's son, John, as acting district secretary-treasurer to complete the unexpired term of another person. The salary for the position was raised $2,000 to $16,000.[38] Lippi delivered a convention-ending oration that brought a standing ovation. The gathering adopted a resolution praising their president's four decades of "unselfish dedication and devotion" to the union's cause.[39] However, the *Scranton Times* characterized the event as a "nauseating performance" by the "once honorable and respected UMWA."[40]

The incumbent president had little trouble winning another four year term in 1965. Within a few months, however, he was on his way to the Lewisburg penitentiary. Lester Thomas, a union official from Harrisburg, was appointed as temporary district president.[41]

Throughout his incarceration Lippi retained a strong desire to control District 1. He decided to position himself for the next convention in December 1968, when his probation would be at hand. Loyal allies initiated a reelection campaign in the face of opposition from some union members who were critical of Lippi's criminal record. The *Wall Street Journal* published a story on Lippi's candidacy, wondering whether the

Fig. 47. UMWA District 1 President August J. Lippi refused to testify at one state initiated hearing. He was later convicted of income tax and bank fraud charges and sent to Lewisburg federal prison. (Courtesy of *Times Leader*)

anthracite miners of northeastern Pennsylvania would "Re-elect [prisoner] No. 863245?"[42]

Word of Lippi's candidacy mobilized officials at the U.S. Department of Labor. They insisted that his convictions precluded office-holding for five years following the conclusion of his sentence.[43] The department also pressured the union's International Executive Board (IEB) to halt Lippi's designs. In early August 1968, the IEB met in Wilkes-Barre and placed District 1 under trusteeship. A provisional government appointed by the union's Washington office would conduct the district's affairs.[44]

Lippi was released from prison on March 21, 1969, after serving forty-one months. He returned home to Exeter twenty pounds leaner and sickly. Questioned about his future plans, he replied that doctors recommended rest because of his failing health. He and his wife soon filed for bankruptcy and the government wrote off as uncollectable his income tax bill of $327,334, plus his $10,000 fine for bank fraud.[45]

On May 10, 1970, at age sixty-nine, August J. Lippi died.[46] He never served jail time for ownership in the Knox Coal Company, complicity in the Knox disaster, or accepting bribes for labor peace. Bank embezzlement and tax evasion eventually brought him down. His indictments, trials, and convictions, as well as those of other District 1 officials, destroyed the credibility of the UMWA in the northern anthracite area.

The proven corporate and union dishonesty only compounded the community's distress over the River Slope calamity. What had happened to the once powerful anthracite industry? How could the region's bulwark of organized labor have become so contaminated? The river had been breached but so was the collective trust in an industry and a labor organization that had sustained the local economy for more than a century.

However, the problem was not just with the Knox Coal Company. Federal and state grand jury investigations into criminal activities in the entire northern field's industry led to indictments against a total of twenty-two individuals and four coal companies including the above mentioned cases. The range of charges included manslaughter, conspiracy, personal income tax evasion, corporate income tax evasion, bank fraud, and violations of labor and mine safety laws. Twelve persons and three companies were eventually convicted of wrongdoing. Of the twenty-two indicted, nine were officials within the UMWA's District 1 and, of these, seven were found guilty. Six coal company officers were convicted of income tax or labor law crimes.[47]

Numerous other trials resulted from the Knox disaster. Most of the victims' families sued and were given modest settlements after years of delays.[48] Dougherty's Peeley Coal Company was successfully taken to court for failing to pay overtime wages. The Commonwealth sued the Knox and the Pennsylvania Coal Companies for damages caused by the disaster but the former declared bankruptcy and the latter successfully argued against its liability. A few small coal businesses unsuccessfully sued the Knox and the Pennsylvania for damages.

However, the legal system never actually found any individuals or organizations guilty for causing the events of January 22, 1959. The seven persons eventually convicted for wrongdoing in the Knox case—Dougherty, Fabrizio, Lippi, Sciandra, Alaimo, Piasecki, and Argo—were apprehended on labor law or income tax violations. Their penalties, moreover, were relatively light.

WHY DID THE KNOX DISASTER HAPPEN?

The four investigations, as well as several other written accounts over the years, have tried to address the question: How could people with years of mining experience have so endangered human lives, their own companies, and the future of an entire industry? The short answer to the question, undertaken by virtually all investigative and journalistic accounts, has focused on greedy or incompetent individuals, and/or a "chain of events" that, once set in motion, seemed to automatically lead to disaster. For example, one editorialist referred to the Knox case as a "greedy blunder" by corrupt individuals.[49] Indeed, survivors and citizens to the present express indignation over the rapacity that led to the tragedy.

The Joint Committee of the Pennsylvania Legislature, on the other hand, in its "chain of events" explanation, took the discussion one step further when it exposed larger-level causes such as inadequate training, lack of regulations in coal mine leasing, outdated mining laws, and Knox's incentive pay system that encouraged careless mining.[50]

However, another set of causes was equally important. They can be thought of as constituting the long answer to question. The focus here is not on individual greed or negligence but on the machinations of four groups: the large coal corporations, the small independent contractors and lessees, the UMWA, and organized crime. The ties and interactions between these groups are crucial to understanding not only why the Knox disaster occurred but why the deep mining era ended in the northern field.[51]

The story begins with a crucial transformation in the field during the decades preceding the Knox crisis. The change involved a restructuring of production away from the large companies that controlled the mineral rights toward relatively small firms who contracted and leased those rights. When anthracite mining began in the early nineteenth century, the typical business was small, family owned, and entrepreneurial. After 1850 larger companies began buying smaller ones, in a trend that gained momentum after the Civil War. Yet another wave of consolidation occurred at the turn of the century and, by 1900 a handful of large corporations that were parts of even larger financial and railroad conglomerates controlled the northern industry.

During the second decade of the twentieth century, however, in response to the high cost of capital investment, wage demands by their workers, and market pressures from other fuels, the large anthracite producers sought to drastically reduce costs and still boost production. To accomplish these goals they instituted a fundamental change in their method of operation. They began contracting (what today would be called subcontracting) sections of their mines to independent entrepreneurs. The Pennsylvania Coal Company was the leader in fostering the contracting system in the 1920s. The independent contractors hired and paid relatively small work crews of up to twenty or thirty men.

Many, though not all, contractors made their profits by violating the union wage rate, by changing work rules, by taking coal in off-limit areas, and by using unsafe mining practices. Eventually, various forms of corruption permeated the contracting system, including monetary "kickbacks" to company bosses, falsification of coal weights, payoffs of inspectors, and even phantom employees whose wages were collected by bosses.[52]

In the early 1920s, labor leader Rinaldo Cappellini stood at the head of a powerful insurgent union movement against the Pennsylvania Coal Company's contracting system. Most of the company's ten thousand employees complained that the contractors reduced them to a form of serf-

Fig. 48. E. Stewart Milner, former vice president, Pennsylvania Coal Company standing next to his collection of company maps (1992). (Photo by Robert P. Wolensky)

dom similar to the Italian *padrone* system where workers held few rights and remained bonded to an overseer. Moreover, the insurgents, many of whom were recent Italian immigrants, argued that the UMWA provided insufficient protection against the new practice. The violent, strike-ridden campaign to eliminate contracting at the Pennsylvania's collieries proved successful. The movement catapulted Cappellini into the presidency of District 1 in 1923. However, by 1928 the system had returned to the Ewen Colliery and each of the other Pennsylvania Coal Company operations whereupon miners re-ignited the conflict.[53]

In the late 1920s and early 1930s, with the same company again leading the way, the large firms went one step further and began leasing entire mines and even collieries (the mines plus all of the surface buildings and equipment) to independent operators. As a result, the Pennsylvania Coal Company virtually went out of the business of mining its own coal. Between 1939 and 1959, for example, the firm issued leases to an average of forty-four independent companies per year. These lessees together produced an annual average of over 1.5 million gross tons of coal worth millions of dollars.[54] The lessor received a handsome royalty for each ton mined, amounting to millions more. The other major companies—the Lehigh Valley, Hudson, Glen Alden and Susquehanna—also moved into the contracting and leasing strategy, though on a smaller scale at first. Miners who had formerly worked for the large firms now had to toil for the lessees.[55]

Workers continued to protest against the new system throughout the late 1920s and early 1930s. They added another demand: "equalization," or the sharing of work hours because so many collieries had been closed or worked irregularly due to slack demand caused by the national economic depression.[56] They staged wildcat strikes and walkouts, many marked by violence. The remonstrations culminated in the formation of an alternate union called the United Anthracite Miners of Pennsylvania (UAMP), under the leadership of Thomas Maloney of Wilkes-Barre.

Maloney, a mine worker and the president of a union local at Glen Alden's Stanton Colliery in Wilkes-Barre, despised the system but disapproved of the violence. After leading several strikes he sought a negotiated solution. He was assisted and supported by Msgr. John J. Curran, the seventy-five year-old "labor priest" who had been a traditional supporter of anthracite mineworkers since the days when he helped President Theodore Roosevelt settle a significant strike in 1902. In 1934, Maloney and Curran traveled to Washington to speak with members of Franklin Roosevelt's administration about the difficult situation in the northern field. However, UMWA president John L. Lewis had risen to a position of influence within the administration and on the National Labor Board so the insurgents received no real hearing. Lewis and the

UMWA successfully worked with the operators and even the national administration to destroy the upstart organization. The UAMP had been defeated by the fall of 1935.

Tragically, on Good Friday, March 5, 1936, Maloney, his five-year-old son, and another person unrelated to the mining industry were murdered by package bombs. The explosive devices apparently had been sent by a disgruntled member of the defeated union. Maloney's death marked the end of organized insurgency movements in the anthracite fields.[57]

With the dual union movement beaten, the contract-leasing system expanded. It became so widespread that by 1942 the northern field could claim 104 separate coal companies, virtually all of them independents working under contracts and leases. As late as 1952, with the industry still in decline, the number stood at seventy-six.[58]

The Pennsylvania Coal Company's dealings with the Knox company clearly illustrated the pattern. In 1943 the firms agreed on a lease for the Schooley workings. Sixteen leases and adjustments to leases followed between 1943 and 1958. Among them was a supplemental lease issued by the Pennsylvania in 1954 allowing Knox the right to mine the Pittston Vein, and thereby dig the River Slope. A subsequent lease agreement permitted the slope's extension into the fateful area under the Susquehanna River.

BENEFITS OF THE CONTRACT-LEASING SYSTEM

The contract-leasing system presented a significant opportunity to small entrepreneurs which allowed them to get started in the anthracite business. A handful of them even grew into large, million-dollar companies with control over entire collieries. The contractors and lessees were only too happy to take the work and the lessors were only too happy to offer it. The system brought even greater benefits to the larger firms. It lowered costs—especially labor costs—as one independent was forced to outbid another, usually at the expense of safety and the UMWA wage agreement. It also allowed the lessor to retain control over key aspects of the coal business while avoiding the actual mining. Initially, the usual arrangements required the lessee to deliver its coal to the lessor. The lessor would then process the coal through its breakers, transport it over the railroad with which it had a contract and, finally, the lessor would sell the coal. Consequently, a company like the Pennsylvania could dictate the terms of an agreement in order to control processing, transportation, and sales, while avoiding the actual mining. Why avoid mining? Because it entailed the greatest risk, the highest costs, and the largest potential for labor conflict.

Regarding the latter, the companies had a long history of rancorous relationships with workers. Labor constituted about 75 percent of the operating cost of a typical anthracite enterprise.[59] For three-quarters of a century, the large companies had fought with workers and their unions over wages, rules, and other work place issues. Labor-management relations in the anthracite were among the most conflictual of any American industry. Therefore, perhaps most importantly, the contract-leasing system brought "discipline" to a traditionally militant workforce and eventually weakened (and corrupted) the principal labor union. This is to say that the independents accomplished what the large companies could not: tame the workers and in effect break the anthracite miners' union.[60] According to William Graham, the last chief executive of the Penn Anthracite Collieries Company, a mid-sized operator which had decided to contract and lease all of its holdings in Scranton:

> These people [lessees] could do what we couldn't do. . . . They could do things that the company wouldn't, couldn't do . . . such as get the extra car of coal from the men which often meant the difference between profit or no profit.
>
> The bigger coal companies had the union, it was there and they were on your back and you paid the prevailing wages. These [lessees] were like [alcohol] bootleggers, the rum runners or the whiskey guys that they made the whiskey, had their stills. These [lessee] fellows had their little mines. They wanted to do what the hell they wanted to do, because they were individuals, I guess. Of course, the union didn't want that, they wanted everybody to be [union]. So I think there had to be a little bit [conflict]. Somebody could get somebody to work for them for less money than a union rate.[61]

By 1940 a new era in deep mining had emerged in the northern field. Dominated by the contract-leasing system, it was characterized by a corporate strategy designed to squeeze profits out of existing mines without resorting to heavy capital investments. It marked a labor relations philosophy that favored union busting and output over workplace cooperation and safety. The operators, therefore, responded to their declining competitive position by transferring their mineral rights to subcontractors who cheapened the workforce, undermined the union contract, but boosted production. It was a bargain that worked for a time but in the end extracted an extremely high price.

ORGANIZED CRIME ENTERS THE PICTURE

To complicate matters, individuals with reputations as members of organized crime secured many (though by no means all) contracts and leases. Nowhere was the pattern more prevalent than at the Pennsylvania Coal Company. For example, Santo Volpe, called "the first boss of the Northeastern Pennsylvania crime family" by the Pennsylvania Crime Commission, obtained fifty-six leases and lease supplements between 1934 and 1948. Volpe began as a contractor in the early 1920s and by the 1940s his Volpe Coal Company held leases on three large collieries. John Sciandra, named by the Crime Commission as the top boss after the retirement of Volpe, secured seven leases as a partner in the Saporito Coal Company between March 21, 1939 and January 27, 1943, and sixteen other leases between 1943 and 1958 as one of the principals in the Knox Coal Company.[62] Numerous other individuals with criminal reputations maintained similar agreements with the Pennsylvania and each of the other large companies.

Did the Pennsylvania Coal Company realize to whom it was leasing? E. Stewart Milner, one of the firm's last surviving officers, said that people in his company did know that they were leasing to individuals with criminal reputations, but no one asked questions:

> I never heard any comment on it. No one really commented on it. I suppose it was probably well known throughout the region and everything, but you just kept your mouth shut or you were in big trouble.[63]

Milner rationalized the dealings between his company and criminal characters as strictly "a business relationship."

Organized crime also gained influence at the highest ranks of District No. 1, UMWA. According to the Crime Commission's *1980 Report*:

> Members of the organized crime family gradually took

over both coal companies and United Mine Workers Union locals and benefited from the "sweetheart" contracts that they were able to obtain.⁶⁴

August J. Lippi's illegal partnership with John Sciandra in the Knox Coal Company illustrated organized crime's influence over the district office. Dominick Alaimo's presence at the Apalachin mob meeting and his leadership position in Local 8005 further demonstrated organized crime's ties to the district. Of course, organized criminals during this period were well-known to have infiltrated and corrupted numerous American labor unions, a taint from which all of organized labor still suffers.⁶⁵

The long answer to the Knox disaster, therefore, leads to the shady dealings between the four principal parties who controlled the twentieth century anthracite business. The only significant group left out was the ordinary mineworkers, who at first fought the perversion but in the final analysis stood defeated. Their plight was made worse by the significant drop in mining employment between 1925 and 1940, causing tens of thousands to leave the coal fields in search of work in New York, New Jersey, Maryland, as well as New England and the Midwest. For those who remained behind, the options in mining employment were few. The case of Knox miner John Gadomski was typical:

> I went in the service for two years. I served in the Pacific theater in the navy. I was with the Seabees. I was a cook and baker on Okinawa with a battalion. After two years I had enough points to get out and I come home, in 1946, May 1946. There was nothing [no work] around here. Only the mines were working.
>
> So November of '46 I went to the Payne Coal Company in Exeter. I went into the service when I was seventeen and I came out when I was just nineteen years old.
>
> I worked for the Payne Coal Company until they closed down. They sold it to Capone Coal Company. Capone went bankrupt so I went with another company, Fabrizio. I worked for him for a while and he went bankrupt in a couple years. So from there I went I got a job with Bernardi up on top of the hill here. That's a little mine shaft. A dog hole. Oh boy, all pitches, anywhere from twenty degrees to forty-five degree pitches. [We were] robbing [pillars] up there. I saw one

guy killed there. He was removing cinder blocks from under an airway and a boulder came down and crushed him. We had to pry him out, jack him up.

Safety went to the wind then. Everything went to the wind. You just got the coal that's what you got paid for. You didn't get paid for being there all day, you got paid for how much coal you loaded. If you didn't load no coal you didn't get paid. . . .

And from there Bob Dougherty who owned the Knox approached me. He said, "John, I got a good place for you." He said, "I know you are a good worker. I have the results. Could you get a crew together?" I said, " Yea, I got a crew. The one I got here." He said, "Good, start the work in the morning." So I ran over there. Beautiful coal. Anywhere from three feet to six feet high all clean coal. We worked in the bottom Marcy.[66]

Moreover, mine inspectors who represented the front line of governmental regulation often turned away from their responsibilities. George Gushanas, former superintendent at the Glen Alden's Huber Colliery, said that inspectors were in a precarious situation:

If you're working [inspecting] a company and they're making big bucks and you tell them, "Hey, we got to stop making them big bucks," what do you think is gonna happen? You're looking for a job.[67]

In the worst cases, inspectors were bought off. Joe Costa worked for one company where the owner gave him $100 every other month to distract the inspector:

They [the inspectors] didn't do nothing. It was a shame. They didn't even go inside, that's what really killed me. Never went into the mines. Just filled out the papers. I remember [lessee's name] said to me, "Joe, here's a hundred dollars. Take the mine inspector out and show him a good time. So I took him fishing and we went out."

I used to get a hundred dollars or so and take the mine inspectors out for dinner and wine and dine them about every other month. They would only inspect around the outside of the mine if we kept them happy. It was a payoff for them.[68]

In summary, the contract-leasing system is crucial to understanding an important, though often unappreciated legacy of the anthracite era—a culture of corruption. While the industry had long been known for perfidy, under this new mode of operation organized criminals—some members of syndicates, others not—took the deception to new levels.[69]

Owners, bosses, inspectors, and mineworkers alike knew that illegalities had become epidemic. They realized that powerful corporations with Wall Street addresses used contractors and lessees to accomplish questionable or illegal goals. They knew that organized crime had become part of the cancer. Finally, they knew that the UMWA had turned away from the mineworkers to become an accomplice in the scandal. Many otherwise upstanding citizens participated in the crooked dealings. The culture of corruption that had engulfed the industry caused serious damage to the community's social and moral fabric, leaving wounds that remain to the present.

THE FINAL CHAPTER ON THE ANTHRACITE INDUSTRY

As serious as it was, the Knox disaster did not, by itself, bring an end to anthracite mining in the northern field. Indeed, by early summer 1959, the massive pumping campaign paid for by the government had lifted more than eleven billion gallons of water from the earth, approximately equal to the amount that had flowed underground following the Knox rupture. Although millions of gallons of make water continued flowing underground, the relatively low pool levels could have been maintained and numerous mines could have been rehabilitated. But the large capital investments required to restore the mines, coupled with the high costs of pumping, discouraged the large companies after the government withdrew from pumping in July 1959. Furthermore, by this time, decades of inadequate investments in advanced technology for mining and burning coal (home heating had always been anthracite's main market) had become readily apparent. On the other hand, the significant technological advances in the oil and natural gas industries meant that anthracite had become terribly uncompetitive.

Within several months of the disaster, the Pennsylvania Coal Company and the Lehigh Valley Coal Company announced their withdrawals from the anthracite business. Glen Alden's purchase of the Hudson Coal

Company in the spring of 1959 left only two of the large companies—Glen Alden and Susquehanna. Meanwhile, independents such as Louis Pagnotti Sr. and his nephew and partner, James Tedesco, continued to expand their holdings by buying properties from the retreating large companies.[70] In Scranton, Glen Alden sold all of its holdings to the Moffat Coal Company, one of the last firms to mine in the Lackawanna Valley.

Some deep mining continued in the southern end of the Wyoming Valley into the early 1970s by the Blue Coal Company which bought the Glen Alden in 1966. When Susquehanna Collieries went out of business, Blue Coal stood as the field's only very large company. However, it too succumbed to the same problem that had plagued the industry for much of the twentieth century: corruption. The main investors in Blue Coal were alleged organized criminals and their associates, including James Durkin, a former state police trooper and principal owner of Pocono Downs racetrack, and James R. Hoffa, former Teamsters Union president who was a secret stockholder. The principals purchased the company with the intention of shutting down its coal operations, selling off its lands and equipment, and dumping its employees on to the state unemployment insurance system. The contorted workings at Blue Coal truly stand as the final chapter in the northern field's 150-year deep mining history, a debacle fully exposed in a Pulitzer Prize winning five-part investigative report by journalists Gil Gaul and Elliot Jaspin of the *Pottsville Republican*.[71]

With the deep mining era gone, Pagnotti Enterprises has presently emerged as the largest producer in the area, securing virtually all of its product by strip mining. Pagnotti gets the vast majority of its coal from open pits in the middle field.[71] Blue Coal remained in bankruptcy court and other legal entanglements for years, most of its lands recently becoming the property of Earth Conservancy, Inc., a local non-profit organization founded and spearheaded by U.S. Congressman Paul Kanjorski. Earth Conservancy has been charged with rehabilitating the scarred lands for preservation and development. Its operations represent part of the continuing effort, begun decades earlier, to rebuild and remake the northern field's economy in the post-anthracite era.

Fig. 49. The Susquehanna River at Port Griffith today, at the site of the Knox Mine Disaster. (Photo by Robert P. Wolensky)

NOTES TO CHAPTER FOUR

1. Al Kanaar, taped interview, October 27, 1988, WVOHP, tape 1, side 2.

2. "River Still Pouring into Mine," *Sunday Independent*, March 1, 1959, sec. 1, p. 1, 2. Rep. Kopriver chaired the Joint Legislative Committee to Investigate the Knox Mine Disaster.

3. "Many Questions Unanswered in Third Week of Disaster That Trapped 12 Men at Knox," *Times Leader, The Evening News*, February 7, 1959, 1.

4. A so-called "sweetheart contract" involves the corruption of a collectively bargained labor agreement. In the typical case, management offers and a union official accepts special compensation or other rewards in return for lax or non-enforcement of the agreement. It is, in effect, a conspiracy to circumvent a legally binding document.

5. "What Goes On In Local Mining?" (editorial), *Times Leader, The Evening News*, January 24, 1959, 8.

6. The fatality statistics are from the Pennsylvania Department of Mines and Mineral Industries, *1961 Annual Report,* Anthracite Division, p. 10. Anthony F. C. Wallace makes this point in S*t. Clair: A Nineteenth-Century Coal Town's Experience With A Disaster-Prone Industry* (New York: Knopf, 1987), especially chapter 5.

7. "State Bill Calls for Inquiry of Disaster," *Times Leader, The Evening News,* January 28, 1959, 3. The Senate members of the Joint Committee included Frank Kopriver, Jr. (chairman, Allegheny); Harold E. Flack (Luzerne County); Paul H. Mahady (Westmoreland County); Martin L. Murray (Luzerne County); Paul L. Wagner (Schuylkill County). The House members were James Musto (vice chair; Luzerne County); Laurence V. Gibb (Allegheny County); James J. Jump (Luzerne County); Stanley A. Meholchick (Luzerne County) William J. Reidenbach (Lackawanna County). Attorneys John F. Fullerton of Harrisburg and Sidney L. Weinstein of Philadelphia served as counsel. For technical advice the committee relied on consultants F. Edgar Kudlich of Kingston, Dr. D. R. Mitchell of The Pennsylvania State University, and Dr. H. L. Hartman. The committee's final report was published as *Report Of The Joint Committee To Investigate The Knox Mine Disaster,* July 27, 1959. F. Edgar Kudlich, an engineering consultant and professor at Penn State University wrote a separate report for the committee, *Confidential Report On The Technical Aspects On The Knox Disaster of January 22, 1959 And Recommendations For Legislation,* May 18, 1959.

8. "Fabrizio, Dougherty Ordered to Appear Before State Senate: They Face Quizzing About Knox Disaster," *Scranton Times,* April 21, 1959, 1, 18. This story reported that the only other instance where citizens were called before the legislature occurred in 1757 when two men were summoned for failing to apologize for published remarks about the General Assembly.

9. *Report of the Joint Legislative Committee,* July 27, 1959, 8.

10. *Report of the Joint Legislative Committee,* July 27, 1959, 72-73.

11. "All Figured in Probe of 12 Deaths: DA Asserts Cases Against Them to Go to Next Grand Jury," *Scranton Times,* August 18, 1959, 1, 14; see also "Knox Report 'Surprisingly Mild'; Legislators Sidestep Fixing Responsibility," *Times Leader, The Evening News,* July 28, 1959, 1, 3. Under Aston's direction, seven indictments were later issued to officials from the Knox and Pennsylvania Coal Companies.

12. The Knox disaster precipitated numerous legislative initiatives to reform the state mining laws. Six bills were introduced into the legislature proposing various measures, the most important being a prohibition of the mining under rivers, streams, and waterways without prior approval of Department of Mines.

13. The report was authored by the three-person commission consisting of Thomas M. Beaney, Willard G. Ward, and John D. Edwards, *Report of the Pennsylvania Commission of Mine Inspectors* (Harrisburg: Pennsylvania Department of Mines and Mineral Industries, April 7, 1959). Warren L. Shirey, the inspector at the River Slope when the disaster occurred, was cited as having been appointed to the commission by Secretary Joseph T. Kennedy in "Mining Commission To Probe Disaster," *Times Leader, The Evening News*, January 25, 1959, 4; however, the appointment was withdrawn because of some controversy surrounding Shirey's inspections at the River Slope.

14. William Rachunis and Gerald W. Fortney, *Report of Major Mine Inundation Disaster, River Slope Mine, May Shaft Section, Schooley Colliery, Knox Coal Company, Incorporated, Port Griffith, Luzerne County, Pennsylvania* (Wilkes-Barre: U.S. Bureau of Mines, 1959), 15.

15. "2 U.S. Agencies Probe Tieup of Mob to Mines: Jury May Get Knox Records," *Scranton Times*, March 3, 1959, 3.

16. *1980 Report: A Decade of Organized Crime* (Harrisburg: Pennsylvania Crime Commission, 1980), 59.

17. On the historic Apalachin meeting see *Interim Report of the* [New York] *Joint Legislative Committee on Government Operations, on The Gangland Meeting In Apalachin, New York*, June 25, 1958; also Steven Fox, *Blood and Power: Organized Crime in Twentieth Century America* (New York: Penguin, 1989) 326-29, 337; and Joseph Bonanno, *A Man Of Honor: The Autobiography of Joseph Bonanno* (New York: Simon and Schuster, 1983), 207-216.

According to the Pennsylvania Crime Commission's *1970 Report On Organized Crime*, the other three local Apalachin participants were Russell Bufalino, the alleged head of organized crime in northeastern Pennsylvania, Angelo Joseph Sciandra, whose mother was a major stockholder in the Knox Coal Company, and James Osticco. The Special Group's national offices were located in Los Angeles, Miami, Chicago, and New York. (See "Antiracket Unit Planned by U.S.," *Scranton Times*, April 11,

1959, 3.) Alaimo, Bufalino as well as the others, were heavily invested in the region's non-union garment industry.

18. The Pittston area had become known as the regional center of organized crime. See "Week's Review of Activities Following Knox Mine Disaster," *Sunday Independent,* February 22, 1959, sec. 1, p. 2.

19. See "Union Aide Named In Mine Pay-Offs: Apalachin Figure Indicted With Officials of Coal Company in U.S. Drive," *The New York Times,* March 4, 1959, 16; "Jury Gets Alaimo Case Today '58 Testimony Seen as Factor: Daring Move by U.S. Jolt to Defendant," *Scranton Times,* May 19, 1959, 1, 18.

The Pennsylvania Crime Commission's *1980 Report* listed Alaimo as a soldier in the Russell Bufalino crime family, and cited "labor racketeering" as one of his criminal credentials. In a personal, untaped interview on August 8, 1989, Alaimo denied any criminal complicity in the Knox corruption and, furthermore, characterized the disaster as "an act of God."

20. Three Local No. 8005 officials were not indicted: Barney Petcovyat, committeeman; John Pientka, vice president; Williard Flynn, secretary.

21. When Washington reporters asked UMWA President John L. Lewis and Vice President Thomas Kennedy about Lippi's indictment as well as his status within District 1, they replied "no comment." See "Mine Racket Expose to Continue: Indictment of Lippi Opens 'New Avenues,' Declares Prosecutor," *Times Leader, The Evening News,* August 7, 1959, 1, 3.

22. *U.S. v. Louis Fabrizio,* Crim. A. No. 1251, U.S. District Court, Delaware, April 19, 1961 (*Federal Supplement,* vol. 193, St. Paul: West Publishing Co., 1961, 446-50); "Fabrizio Wins Acquittal on Taft Act Charges: Judge Denies Similar Dougherty Motion, Affirms Lippi Ruling," *Times Leader, The Evening News,* April 20, 1961, 1, 3.

23. "Union Leader Guilty of Bribery," *The New York Times,* June 21, 1960, 27. Two decisions were rendered on the new trial. See United States District Court, District of Delaware 190 F. Supp. 604; 47 L.R.R.M. 2537; 42 Lab. Cas. (CCH) P16,792, February 2, 1961; *United States of America v. August J. Lippi,* Crim. A. No. 1269, U.S. District Court, District of Delaware, 190 F. Supp. 604, 47 L.R.R. M. 2537; 42 Lab. Cas. (CCH) P. 16, 792, April 19, 1961. See also "Lippi Wins 2nd Time," *Times Leader, The Evening News,* February 7, 1961, 1; "Fabrizio Wins Acquittal on Taft

Act Charges: Judge Denies Similar Dougherty Motion, Affirms Lippi Ruling," *Times Leader, The Evening News*, April 20, 1961, 1, 3.

24. "Ex-union Leader Jailed," *The New York Times*, March 11, 1961, 10; "Piasecki, Argo Barred From Office in UMWA," *Scranton Times*, May 22, 1959, 1, 15.

25. "2 Freed in Mine Deaths," *The New York Times*, July 12, 1960, 24.

26. *Commonwealth of Pennsylvania, Appellant, v. Louis Fabrizio, Robert L. Dougherty and August J. Lippi*, The Superior Court of Pennsylvania 197 Pa. Super. 45; 176 A.2nd 142, December 15, 1961; see also "Convictions Voided in Mine Death of 12," *The New York Times*, February 8, 1961, 22; "Knox Mine Convictions Upset: Manslaughter Charge and Conspiracy Case Ruled Out by Court; Lippi, Fabrizio, Dougherty Freed," *Times Leader, The Evening News*, February 7, 1961, 1, 3; "Court Asked to Reinstate Knox Mine Charges: Superior Bench Urged to Upset Dismissal of Conspiracy Charges," *Times Leader, The Evening News*, June 13, 1961, 1.

27. Stephen A. Teller, taped interview, June 27, 1990, WVOHP, tape 1, side 1. Another key piece of evidence linking August J. Lippi as an owner of the Knox Coal Company was a life insurance policy purchased by and indemnifying the company in the event of Lippi's death.

28. The strategy was discussed by Teller in his taped interviews, June 27, 1990 and January 5, 1995, WVOHP; and by Degillio in his taped interview July 30, 1990, WVOHP. Teller prosecuted the cases against Groves and Receski. Degillio served as defense counsel for Receski and for Josephine Sciandra in her income tax evasion case.

29. "Dougherty Joins Fabrizio in Federal Penitentiary," *Wilkes-Barre Record*, May 7, 1964, 16

30. Dougherty's ten character witnesses included some very prestigious community leaders: Dr. Leonard Corgan, a mine operator; Con McCole, former mayor of Wilkes-Barre; Thomas Barry of the *Sunday Independent*; Charles Connelly, a former state police officer; Francis Crossin, state representative; Joseph Conahan, mayor of Hazleton; and Msgr. Francis A. Costello of the Diocese of Scranton who sent a letter to the court citing his fifty-year friendship with the accused. See "Dougherty Gets Year; Fined $10,000: Coal Mine Owner Must Serve Term in Federal Prison," *Times Leader, The Evening News*, May 6, 1964, 1, 3.

31. "Tax Cases Are Joined," *Times Leader, The Evening News*, Sep-

tember 21, 1963, 3.

32. Ed E. Rogers, "His Return To District Is Ordered: Miners' Chief Seized On Airliner Bound For South America," *Scranton Times*, December 21, 1962, 3, 26.

33. See Norman Daileda, taped interview, WVOHP, May 28, 1998, tape 1, side 1; and *U.S. v George J. Daileda and August J. Lippi*, U.S. Court of Appeals Third Circuit, 342 F.2d 218, argued January 22, 1965; decided March 9, 1965. George Daileda died of natural causes before being sentenced.

34. See "Lippi May Stage Last Fight to Escape Jail," *Scranton Tribune*, November 4, 1965, B-1; "Lippi Starts Serving Term in Prison Today," *Scranton Tribune*, November 5, 1965, B-1; "Serves 41 Months of 5 Year Term," *Times Leader, The Evening News*, March 21, 1969, 6.

35. "Shhh," *Times Leader, The Evening News*, October 10, 1959, 9.

36. "Union Awaits Own Probe in Lippi Case," *Times Leader, The Evening News* August 10, 1959, 2; "Lippi and Staff Unopposed for New Terms in Mine Union," *Times Leader, The Evening News*, August 30 1960, 1; "UMW Not to Prevent New Term for Lippi," *Wilkes-Barre Record*, August 31, 1960, 9.

37. "Union Votes Lippi, Son Large Pay Increases," *Scranton Times*, March 25, 1965, A-3. District 1's membership stood at 5,814 in 1960 but declined to 4,055 by 1965. See "UMWA Rolls Decline by 30%," *Scranton Times*, March 22, 1965, D-1.

38. Lippi wrote a letter thanking Boyle for his son's appointment: "I cannot find words to convey to you my deep sense of appreciation. I hope soon to try to express my thanks to you in person." He added: "Needless to say, you will continue to have my unswerving support in everything you undertake. Perhaps, I may be able in some small way to repay you for your most recent display of thoughtfulness." See August J. Lippi to President W.A. Boyle, March 11, 1965, August J. Lippi File, UMWA Papers, Labor History Archives, Penn State University, Correspondence-Personnel, January-May, 1923-1974, No. 3/M 10, T4.

39. *Proceedings*, Sixth Quadrennial Constitutional Convention of District No. 1, UMWA, March 22-24, 1965, Wilkes-Barre, Pa., 232.

40. "Nauseating Performance," *Scranton Times*, March 26, 1965, 9.

41. "Lippi is Removed to Lewisburg to Begin His Five Year Sentence," *Scranton Times*, November 5, 1965, 3, 6; "UMWA Picks Thomas

in Place of Lippi," *Times Leader, The Evening News*, November 1, 1965, 6; "Lippi Replaced as District 1 Head," *Wilkes-Barre Record*, November 1, 1965, 1.

42. "Re-elect No. 863245?" *Wall Street Journal*, August 20, 1968, obtained from UMWA Papers, Labor History Archives, Penn State University, District 1 - District 31, box 2, 1/L12, location: box ZZ/4, folder: District No. 1, 1960.

43. Edward J. Carey [general counsel, UMWA] to President W.A. Boyle, Re: August J. Lippi, May 15, 1967, August J. Lippi file, UMWA Papers, Labor History Archives, Penn State University, Correspondence-Personnel, January-May, 1923-1974, No. 3/M 10, T4.

44. See "UMWA Executive Board Imposes Trusteeship and Establishes Provisional Govt. in District 1," *Wilkes-Barre Record*, August 3, 1968, 2; obtained from UMWA Papers, Labor History Archives, Penn State University, District 1, box 2, 1/L12, July 7, 1968.

45. "Lippi Declares He's Bankrupt," *Scranton Times*, March 23, 1967, 3; "Lippi's $10,000 Fine of 1964 Not Yet Paid," *Scranton Times*, August 20, 1969, 3. Robert Dougherty paid the $10,000 fine resulting from his income tax evasion conviction.

46. "August J. Lippi Dies: Former UMWA Leader," *Times Leader, The Evening News*, May 11, 1970, B-3.

47. The indicted individuals were: Dominick Alaimo, Anthony Argo, Albert Biscontini (part owner of Newport Excavating Co; personal and corporate income tax evasion) Frank Cardoni (assistant to August J. Lippi and board member of UMWA District 1; personal income tax evasion), William Dombrowski (secretary, UMWA Local 7519; taking bribes for labor peace), Robert Dougherty, Louis Fabrizio, former Governor John Fine (part owner of Newport excavating; personal and corporate income tax evasion), Ralph Fries, Robert Groves, Philip Gelso (owner No. 14 Coal Company; payoffs for labor peace), Sam Gelso (owner No. 14 Coal Company; payoffs for labor peace), Thomas Larkin (Anthracite Conciliation Board umpire; personal income tax evasion), August Lippi, Donald Morgan (brother-in-law and tax advisor to John Fine; complicity in income tax evasion), Charles Piasecki, William Receski, Fritz Renner, John Salvo (committeeman, UMWA Local 7519; taking bribes for labor peace); Josephine Sciandra, John Shipula (president UMWA Local 7519; taking bribes for labor peace), Leonard Statkewicz (board member, UMWA

District 1; personal income tax evasion).

The companies indicted were: Avon Coal Company, Newport Excavating, Knox Coal Company, and the Peeley Coal Company.

Only the following individuals were convicted: Alaimo, Argo, Dombrowski, Dougherty, Fabrizio, Philip Gelso, Sam Gelso, Lippi, Piasecki, Salvo, Sciandra, and Shipula. They received sentences ranging from suspended to five years in prison. The three convicted companies were the Avon, the Peeley, and the Knox.

48. See, for example, "3 Widowed by Knox Disaster Get $10,000 Settlements Each," *Times Leader, The Evening News,* July 22, 1966, 3.

49. "Coal Companies Starting Too Late to Care About Public Relations: They Want Something Now," (editorial) *Sunday Independent*, February 14, 1960, sec. 3, p. 6.

50. Knox revolutionized the standard per-car piece rate that had long been established in the region. Traditionally, a typical crew of three workers (one miner and two laborers) produced three cars per man per shift. However, to boost production the company devised an incentive plan whereby crews received a bonus for each extra car produced above the old minimum. According to Dougherty: "A few years [after World War II] we also put another advantage in there for the men by paying them the basic car rate with this addition. If there is three men in the place, they loaded nine cars. The tenth car on that job paid twenty-five cents extra, the eleventh car on that job paid fifty cents extra, up to a point where if they load enough coal maybe the last car of the day that they loaded over the basic car rate would be maybe $3.00 a car extra on the pyramiding system. . . . The men earned between $30 and $50 a day—not a week, a day. (From Dougherty's testimony before the Joint Legislative Committee, April 23, 1959, 1423-1425.) The incentive system encouraged mining gluttony at the expense of safety. It was not uncommon for a Knox work team of three men to produce thirty to fifty cars per shift .

51. The only detailed published accounts of the Knox disaster are by E. W. Roberts, "The End of Anthracite in Wyoming Valley: The Knox Mine Disaster, 1959, in his book *The Breaker Whistle Blows* (Scranton: Anthracite Museum Press, 1984), chapter 11; and George. A. Spohrer, "The Knox Disaster: The Beginning of the End," in *Proceedings and Collections of the Wyoming Historical and Geological Society* 24 (1984): 124-145. Neither discussion explored what has been here called "the long answer."

52. These types of corruption were discussed in oral history interviews with David Panzitta, Joe Costa, William Hastie, Leo Butsavage, and others. Several of the individuals and companies listed in note 47 were indicted and convicted on such charges.

53. On Cappellini and the insurgent movement to eliminate contracting in the 1920s see Perry K. Blatz, *Democratic Miners: Work and Labor Relations in the Anthracite Coal Industry, 1875-1925*, (Albany: State University of New York Press, 1994), 241-250.

54. Pennsylvania Coal Company Papers, Pennsylvania Historical and Museum Commission (PHMC), MG 282, Royalty Records, 1883-1968, Recapitulation of Tonnage and Coal Royalty Receipts by Lessees for Years 1939-1965. A complete listing of the company's leases can be found in he Pennsylvania Coal Company Papers, PHMC, MG 282, contract file index, and in the volumes containing the minutes of the stockholders and executive committee meetings, 1838-1971. A full analysis of the Pennsylvania company and its involvement with the contract-leasing system can be found in the authors' more detailed book on the Knox Mine Disaster, forthcoming from the University of Illinois Press.

55. In addition to a reviewing the Pennsylvania Coal Company leases housed at the Pennsylvania State Archives in Harrisburg, the leases of four other large companies have been reviewed for this study: the Lehigh Valley (eighteen volumes, from 1884 to 1952, housed at Pagnotti Enterprises in Wilkes-Barre, which bought the Lehigh Valley in the 1960s); the records of the Glen Alden (housed at Earth Conservancy, Inc. in Ashley which took over Glen Alden's properties in the 1980s), the records of the Hudson (also housed at Earth Conservancy), and the records of the Lehigh and Wilkes-Barre Coal Company (also housed at Earth Conservancy). It is clear that all of the firms participated widely in the contract-leasing system.

Also reviewed were the leases of one middle-sized company, Penn Anthracite Mining Company, and one small firm, Lacoe and Schiffer Coal Company, a company established do nothing but secure mineral rights and lease them. The authors are indebted to William Graham for the Penn Anthracite information, and to Tony DeAngelo for providing the corporate minute book of the Lacoe and Schiffer company.

56. On the successful pursuit of equalization in the Panther Valley section of the anthracite region see Thomas Dublin, "The Equalization

of Work: An Alternative Vision of Industrial Capitalism in the Anthracite Region of Pennsylvania in the 1930s," In Lance Metz (ed.) *Canal History and Technology Proceedings, 13 (1994): 81*-98.

57. Michael Fugmann, a good friend of Maloney, was convicted of the murders and subsequently executed in the electric chair for the crimes. On Maloney and the UAMP see Douglas K. Monroe, unpublished Ph.D. dissertation (Georgetown University) *A Decade of Turmoil: John L. Lewis and the Anthracite Miners, 1926-1936* (Ann Arbor: University Microfilms, 1977). See also John Bodnar, *Anthracite People: Family, Unions and Work, 1900-1940* (Harrisburg: Pennsylvania Historical and Museum Commission, 1983). Also Chester Brozena, taped interview, WVOHP, December 3, 1988.

Many, if not most local mineworkers, however, have argued that Fugmann was framed. The taped oral history interview with Lewis Casterline expressed this widely accepted view (Lewis Casterline, taped interview, December 12, 1988, WVOHP, tape 2, side 1), as did the account of radical organizer Steve Nelson. see Steve Nelson, James R. Barrett, and Rob Ruck, *Steve Nelson: American Radical* (Pittsburgh: University of Pittsburgh Press, 1981), 171-72.

Margaret Maloney Bednark, at age sixteen, suffered severe wounds in the explosion that killed her father and brother. See her taped interview, June 10, 1998, WVOHP, tape 1, side 1.

58. W.C. Macquown's 1942 and 1952 *Maps of the Anthracite Coal Fields of Northeastern Pennsylvania* (Pittsburgh: National Coal Publications, 1942 and 1952) list each of the firms—lessors as well as lessees—along with a mapping of the mines they worked. Surprisingly, however, the Knox Coal Company was absent from the 1952 list, its operations attributed to the Pennsylvania Coal Co.

59. The 75 percent figure for labor costs was first computed by the Anthracite Coal Strike Commission of 1922. S.D. Warriner, a coal operator who was also Chairman of the Anthracite Operators' Association in 1925, also put the labor cost figure at 75 percent (S.D. Warriner, "Reply of the Operators' [to the Miners' Demands]," in *The Anthracite Strike of 1925-1926* (Philadelphia: The Anthracite Bureau of Information, 1926), 6. Donald L. Miller and Richard E. Sharpless in *The Kingdom of Coal* (Philadelphia: University of Pennsylvania Press, 1985) said it was 70 percent, as did the Hudson Coal Company in *The Story of Anthracite*, 1932.

60. On the long history of rancorous labor-management relations in the anthracite region see Victor Green, *Slavic Community on Strike* (South Bend: University of Notre Dame Press, 1968); Ellis W. Roberts, *The Breaker Whistle Blows: Mining Disasters and Labor Leaders in the Anthracite Region* (Scranton: Anthracite Museum Press, 1984); Donald L. Miller and Richard E. Sharpless, *The Kingdom of Coal: Work, Enterprise, and Ethnic Communities in the Mine Fields* (Philadelphia: University of Pennsylvania Press, 1985); Priscilla Long, *Where The Sun Never Shines: A History of America's Bloody Coal Industry* (New York: Paragon House, 1989); and Perry K. Blatz, *Democratic Miners: Work and Labor Relations in the Anthracite Coal Industry, 1875-1925* (Albany: State University of New York Press, 1994).

61. William Graham, taped interview, July 11, 1995, WVOHP, tape 1, side 2.

62. The positions of Volpe and Sciandra within organized crime are discussed by The Pennsylvania Crime Commission, *1980 Report*, 50. During the 1950s, Volpe became a business partner of former Governor John Fine in the Newport Excavating Co., a company that leased coal properties from the Glen Alden. The venture led to Fine and another partner, Albert Biscontini, being indicted for corporate and personal income tax evasion (see note 47). Volpe had passed away a few years before the indictments were issued, but he was mentioned as one of the co-conspirators along with yet another partner, Lawrence Biscontini, who had also passed away. Fine and Albert Biscontini were found innocent of the personal tax evasion charge while the corporate evasion counts were eventually dropped by the government.

63. Stewart E. Milner, taped interview, August 3, 1992, WVOHP, tape 1, side 2.

64. The Pennsylvania Crime Commission, *1980 Report*, 50.

65. Stephen Fox in *Blood and Power* (New York: Penguin Books, 1989), provided a good account of the labor corruption initiated by organized crime in the 1940s and 1950s.

66. John Gadomski, taped interview, December 22, 1988, WVOHP, tape 1, side 1.

67. George Gushanas, taped interview, WVOHP, tape 2, side 1.

68. Joe Costa, untaped interviews, August 1, 1992 (second paragraph quote) and July 22, 1995 (first paragraph quote).

69. The Pennsylvania Crime Commission conducted a study in the 1970s entitled, *Coal Fraud: Undermining A Vital Resource*, where organized criminals were defined as any "persons who take part in illegal conspiratorial acts with economic gain the ultimate goal." In another report, *Abuses in the Coal Industry*, published in 1983, the Commission wrote that, "Organized crime is much broader than 'The Mob' or 'La Cosa Nostra'. . . . It encompasses the sophisticated white collar criminal who takes, not by force, fear or intimidation, but instead by stealth and manipulation. It involves criminal acts such as theft by deception, deceptive business practices, securities fraud, wire fraud, mail fraud and conspiracy. It includes any concerted activity or unlawful practice designed to effect large financial gain."

70. See James Tedesco, taped interview, October 8, 1996, WVOHP.

71. "Pulitzer," *Pottsville Republican*, April 17, 1979, 1. The five articles on Blue Coal appeared in the *Republican* between June 12 and June 16, 1978.

72. See Louis Pagnotti III, taped interview, January 11, 1998, WVOHP. Pagnotti's Coal Company, near Hazleton, was hit by a strike of its UMWA Local 803 workforce in 1998 over an old issue—contracting, in this case to non-union coal companies. On the strike see "Miners Bracing for the Long Haul," *Times Leader,* May 3, 1998, 1D; and "Jeddo Workers on Strike: No Justice, No Peace, *United Mine Workers Journal*, 109 (May-June 1998): 4-5.

CHAPTER FIVE
REBUILDING THE REGIONAL ECONOMY IN THE POST-ANTHRACITE ERA

In the anthracite region of Pennsylvania the issue has been the economy ever since I can remember. There is nothing else to talk about or anything nearly as relevant to those people's needs than the economy.
Former Pennsylvania Governor George M. Leader, 1995[1]

The Knox Mine Disaster delivered a virtually fatal blow to deep mining in the northern anthracite field. One estimate placed the direct and indirect job loss at 7500 and the payroll deprivation at thirty-two million dollars. Total anthracite production in 1959 dropped by 94,000 tons.

The decline in anthracite had been coming for some time, however. Production peaked at about 100 million tons in 1917 and headed downward thereafter. The prolonged strike of 1925-26 turned home and industrial users from throughout the Northeast—anthracite's main market area—away from the fuel. In most mining towns during the 1930s, three out of every four men were either unemployed or underemployed. Thousands worked two or three days a week. Many desperate families turned to public assistance programs. Some relied on work by mothers and daughters in the garment and silk industries. Numerous miners opened illegal "bootleg" mines and "dogholes" to earn a few dollars. Communist-led unemployment councils were established throughout the hard coal region to provide assistance and advocate new public policies. One council, headquartered in Wilkes-Barre, published the *Anthracite Unemployed Worker* with a readership of five thousand.

Although output slightly rebounded during World War II, the market for hard coal remained weak by historic standards. After the war, total anthracite employment declined to under 50,000 from nearly 100,000 a decade earlier. Out-migration became a fact of life in anthracite towns, a trend that would continue through the coming decades. Between 1940 and 1950, for example, Luzerne County lost 11 percent of its population.[2]

Recognizing their over-reliance on a declining sector of commerce, government and business leaders joined with civic-minded citizens and labor unions to attract new companies. In Scranton, organizers formed the Lackawanna Industrial Fund Enterprise (LIFE) in 1946, which led to a total of seventy-five plant expansions and fifty-five new factories. LIFE helped create some 7,000 new and 7,000 spin-off jobs by the mid-1950s.[3] Soon after World War II, the Wilkes-Barre Chamber of Commerce established a New Industries Committee and an Industrial Development Fund which set out to raise $350,000 for new enterprises. Manufacturers of clothing, textiles, shoes, and cigars became the chief targets. In response to its economic condition, Hazleton, in southern Luzerne County created CAN DO to spearhead economic rejuvenation.

Still, by the early 1950s, Wilkes-Barre and Scranton held the distinction of being the only two urban areas in the United States with unemployment rates above 12 percent. Between 1947 and 1958, mining jobs in Luzerne County fell from over 35,000 to about 10,000. In Lackawanna County the decline was from 14,000 to about 4,000 (see table 2).[4]

During the 1950s, economic development efforts in the greater Wilkes-Barre area included The Committee of 100, a dedicated group of business, labor, civic, and academic leaders. The committee made contacts with industrial concerns urging them to locate facilities in the area. It also raised funds for industrial parks, business relocations, and plant start-up costs. In 1952 in conjunction with the Industrial Development

TABLE 2. EMPLOYMENT IN ANTHRACITE COAL MINING, NORTHERN ANTHRACITE FIELD 1950-PRESENT

Year	Number of Persons Employed Luzerne County[5]	Number of Persons Employed Lackawanna County[6]
1950	34,500	12,000
1952	29,100	10,400
1954	18,100	3,500
1956	11,900	4,100
1958	10,200	4,000
1960	6,200	1,900
1962	4,200	1,000
1964	4,000	1,200
1966	3,200	700
1968	2,400	400
1970	2,100	500
1972	1,600	300
1974	1,200	300
1976	900	300
1978	1,200	200
1980	1,100	*
1982	1,000	*
1984	800	*
1986	800	*
1988	600	*
1990	600	*
1992	700	*
1994	600	*
1996	300	*
1998	400	*

*The number of persons employed in anthracite mining in Lackawanna County from 1980 to the present remained below 200.

Fund, the committee raised $600,000 from local interests (including at least one coal company), to purchase six hundred acres in nearby Mountaintop which became Crestwood Industrial Park. By 1953 Foster Wheeler Corporation, a heavy equipment maker, opened its doors at Crestwood and employed several hundred people. One year later King Fifth Wheel, a manufacturer of truck parts and transportation components, built a large facility in the park.[7]

Local endeavors to reindustrialize in the wake of anthracite's long slide were coupled with similar state and federal endeavors. In 1956 Pennsylvania Governor George M. Leader signed legislation to enact the Pennsylvania Industrial Development Authority (PIDA), the state's first program to assist distressed communities. PIDA, which exists to the present day, was originally designed to provide low-interest second mortgages to businesses setting up or expanding operations in high unemployment areas. The legislation allowed local industrial development committees to arrange commercial mortgage loans for prospective enterprises and for the Commonwealth to provide additional financing for 30 to 40 percent of building costs. Governor Leader also enacted a provision to ensure that loans went only to firms recommended by local authorities in areas with unemployment greater than seven percent.[8] Wilkes-Barre became one of the first communities to take advantage of PIDA when it attracted writing instrument manufacturer Eberhard Faber to Crestwood Industrial Park in 1957.

Between 1956 and 1978, PIDA approved more than 1,300 loans totaling some $488 million. A substantial share of this assistance flowed to the three main anthracite counties of Lackawanna, Luzerne, and Schuylkill. By 1960 PIDA had committed to nine loans within the greater Wilkes-Barre area alone.[9] Thus, the hard coal region became one of the early examples of an economic development policy that would eventually spread throughout Pennsylvania and other parts of the United States, one where public moneys would be used to subsidize private investment for economic development.

According to Governor Leader, the scale of economic distress in the anthracite region, as well as local efforts to address the problem, inspired his administration to develop and enact PIDA:

> In the anthracite region of Pennsylvania the issue has been the economy ever since I can remember. There is noth-

ing else to talk about or anything nearly as relevant to those people's needs than the economy. What happened was, after World War II, the young people had just left that area in droves. It lost a good deal of that generation, at least the male population. Now some of them came back, but a lot of them left. Some went to Washington, others to Harrisburg for the opportunities. The federal government was expanding during those years and they were able to move into a lot of those jobs and they were good jobs. They gave them the kind of security they never could have had in the coal region.

There was a plan called the Scranton Plan and Wilkes-Barre was doing something similar where they were raising money locally to bring industry in. We were deciding what the state could do in that regard. We held a series of public hearings across the state. The first one was in Wilkes-Barre.

I remember very well the first hearing. It was tremendously supported both by the people who were doing industrial development and by the unions, like the International Ladies' Garment Workers' Union, who wanted to see more jobs. People from Scranton and Wilkes-Barre came in and testified as to what the state should do to attract above ground [non-mining] industry.

Frankly, out of those meetings came the Pennsylvania Industrial Development Authority. That area had a great influence on my industrial development policy. They had the best ideas. And, so, I learned from them and we did PIDA, in part, because of the thinking, experiences, and knowledge of people in the anthracite region.[10]

Leader remained painfully aware of the impact of the Knox catastrophe. His term in office had expired only a few weeks before the calamity and he followed the situation closely. To him, Knox epitomized much of what had gone wrong with King Coal:

There is something about the extractive industries that, somehow, exploitation seems to be the only word that applies. They don't seem to care about the hospitals or the churches or the community buildings or even the infrastructure unless it directly affects them. They just never did anything to help the community. They just got in and they took their money and they did absolutely as little as they could to protect the workers from dust, from cave-ins, from anything. They just did the minimum. That's what it was all about. Get in. Get out! Get their money and get out![11]

State and local efforts boosted the area's economy. The number of firms who either relocated to the region or who expanded operations increased from an average of twenty-five in the mid-1940s to fifty by the mid-1960s. About one-third were garment factories, 20 percent were in the metal trades, while food and beverage establishments comprised another six percent. By 1960 PIDA provided assistance to nearly a quarter of them. Relocating firms totaled 140 between 1940-1970. Forty percent came from the New York City area, while another 20 percent relocated from elsewhere in Pennsylvania or the mid-Atlantic region.

At the federal level, Congressman Daniel J. Flood—whose 11th Congressional District included a large portion of the northern anthracite field with the exception of Lackawanna County—worked to secure the participation of the national government. For the better part of the 1950s, Flood had managed to obtain legislation to support communities in economic distress: "When I came to Congress in 1944, my district was exporting one thing—high school graduates," Flood often recalled.[12] His work culminated with President John F. Kennedy's enactment of the Flood-Douglas Area Redevelopment Bill in May 1961. The legislation supplied low- interest loans to private enterprises who relocated or expanded in economically distressed locales. It also funded worker training programs, infrastructure improvements, and plant construction. The U.S. Commerce Department established an Area Redevelopment Administration (ARA) to administer the program. In 1965 the Appalachian Regional Commission assumed the duties of the ARA. Flood helped convince the federal government to include all of the hard coal counties within the official Appalachian region, permitting access to millions of dollars in development assistance. Furthermore, the Economic Development Administration's office for the eastern U.S. was located in Wilkes-Barre. Funding flowed from this program, for example, to plant construction and water system development for Hazleton's new Valmont Industrial Park.

A widely recognized Washington power broker, Flood served as chair

Fig. 50. Hazleton's Valmont Industrial Park. (Courtesy of CAN DO)

of the Labor, Health, Education, and Welfare Subcommittee and vice chair of the Defense Appropriations Subcommittee. These roles meant that two-thirds of the federal budget fell under his influence. He saw to it that northeastern Pennsylvania received more than its share of Washington's spending. Flood worked to ensure that Monroe County's Tobyhanna Army Depot became the East Coast headquarters for refurbishing military communications equipment. He helped obtain the rerouting of Interstate 81— which traverses the Appalachians—through the heart of the hard coal fields. He committed the national government to the construction of a regional veterans' hospital in Wilkes-Barre. He procured public financial support for the construction of the Avoca Airport serving Lackawanna, Luzerne, and neighboring counties. He also vigorously fought for enactment of the 1969 Coal Mine Health and Safety Act which provided over $500 million in compensation to the 25,000 victims of black lung disease residing in his district.

By 1972 total federal military-industrial spending alone in Flood's 11th District amounted to an astonishing $378 million. In the congressman's view, only the government had the resources and the moral obligation to "invest" so large an amount in a depressed area. In the process of anthracite's decline, however, a region that had once been a veritable fountain of capitalistic wealth had become, in large part, a ward of the state. The boom of anthracite's golden age had turned to the bust of an economy that survived only through government intervention.

Flood even led federal attempts to revitalize the dying anthracite industry. In 1977 he oversaw the initiation of an Anthracite Task Force composed of federal, state, and local officials, members of Congress, and industrialists. The group studied the industry and made numerous recommendations to increase production and expand markets. As a result, the Carter administration agreed to create the Office of Anthracite to promote hard coal. However, budgetary constraints closed the office and many of the task force's suggestions were never implemented.[13]

Joseph McDade of Scranton, representing Pennsylvania's 10th Congressional District, has likewise been instrumental in bringing federal dollars to the anthracite region. During his three decade-plus congressional career, which ended in 1999, McDade held numerous positions on powerful House committees. He also worked assiduously to secure the stability of Tobyhanna Army Depot. Moreover, the multi-million dollar Steamtown National Historic Site in downtown Scranton, operated by the U.S. Department of the Interior, would not have become a reality without his dogged advocacy. The congressman also worked to secure a Small Business Development Center at the University of Scranton to assist entrepreneurs and numerous other successful initiatives to transform the economy of his district.[14]

The combined local, state, federal, and private efforts yielded impressive results. About 80,000 new jobs were created between 1960 and 1980 throughout the entire hard coal region. The labor force grew from 180,000 in 1960, to 206,000 in 1970, to 247,000 by 1980. By 1967 over 13,000 new jobs had been created in the greater Wilkes-Barre area alone, comprising an annual payroll of over $40 million. Crestwood Industrial Park, for example, had fourteen new employers—including a $2 million Radio Corporation of America plant—employing 5,000 people. Employment in mining, meanwhile, declined from about 14,000 around the time of the Knox Mine Disaster to fewer than 3,000 by 1980.[15] Nevertheless, during the 1960s and early 1970s, unemployment in the northern anthracite field had begun to reflect the efforts to rejuvenate the area's economy (see table 3).

TABLE 3. PERCENTAGE OF UNEMPLOYMENT IN THE NORTHERN ANTHRACITE FIELD, 1950-1972

Year	Luzerne County[16]	Lackawanna County[17]
1950	9.3	15.0
1952	9.3	16.7
1954	15.3	14.6
1956	13.0	9.6
1958	16.9	15.6
1960	14.0	12.4
1962	11.6	12.4
1964	7.7	8.4
1966	5.4	5.1
1968	4.1	4.4
1970	5.8	6.3
1972	8.8	4.9

Tropical Storm Agnes Devastates the Wyoming Valley

No one could have anticipated the devastating impact—and temporary setback—that would result from Tropical Storm Agnes. Agnes was declared a threat to Eastern parts of the United States on June 17, 1972. After traveling up the Atlantic coast, on June 22 the storm turned westward toward Pennsylvania and stalled over the eastern and central portions of the state. Three days of torrential rain dumped 14 trillion gallons of water over much of the Commonwealth as well as lower portions of New York state. The Susquehanna River flooded numerous communities in New York, Maryland, and Pennsylvania. When the skies cleared, Pennsylvania had suffered $1.5 billion in damage, more than any other state.

Seventy percent of the total losses occurred in Luzerne County. The damages in the Wyoming Valley were estimated at $1 billion. Over 70,000 people evacuated the area when major flooding began on June 23, 1972. Sixteen of the county's municipalities were directly hit. Fifty thousand people were put out of work. Calling Tropical Storm Agnes the most destructive natural disaster in the nation's history, President Richard M. Nixon declared all sixty-seven Pennsylvania counties a disaster area.[18]

Agnes ushered in a level of disaster assistance unmatched in the nation's history. Thanks to Pennsylvania's clout in Washington coupled with the desire of candidate Nixon to win the November presidential election, on August 9, 1972, House and Senate conferees agreed on the Agnes Recovery Act which allocated $1.6 billion in assistance. A few days later the president signed the measure. The program included $5,000 grants for each household or business plus reconstruction loans at 1 percent interest for thirty years. The state government, under the leadership of Governor Milton J. Shapp, allocated an additional $150 million in disaster relief—the most ever offered by any state—which included $40 million in direct cash grants of up to $3,000 per household. The Wyoming Valley's share of total government allocations amounted to over $1.024 billion.

Though the storm's immediate impact brought much of northeastern Pennsylvania—and a fair portion of the state—to its knees, in the longer term Agnes provided an economic boost. In Wilkes-Barre the City Redevelopment Authority undertook four urban renewal projects, replacing many of Public Square's flood ravaged buildings with modern office and retail space. The Luzerne County Redevelopment Authority spent

Fig. 51. Public Square, Wilkes-Barre 1997. (Courtesy of *Times Leader*)

$160 million on similar projects throughout the valley. Cleanup and reconstruction efforts meant that thousands of homeowners in the flood plain had newly refurbished dwellings at minimal personal cost. Businesses, schools, hospitals, churches, libraries, and community establishments likewise rebuilt.

In short, as local residents liked to say, the community was "coming back better than ever." National publications called Wilkes-Barre "the newest old town in America." *The Boston Globe* called the recovery "One of the greatest comebacks in American history." One journalist in *The Wall Street Journal* wrote an article in 1987 entitled, "How a Flood Turned Around a Pennsylvania City's Economy."[19] A blessing in disguise, this "other" flood brought by the Susquehanna River allowed the Wyoming Valley to remain more competitive than it otherwise would have been.

THE PRESENT ECONOMIC SITUATION

During more recent years, the economy of the northern anthracite field, like the economy of the United States generally, has evolved from manufacturing to services. The garment sector of commerce, once a pillar of the post-anthracite economy, has experienced a steady decline due to an erosion of import limitation policies, the overseas flight of companies in search of cheaper labor, and national free trade policies. At its peak in the mid-1960s, garment shops employed over 18,000 persons in Luzerne County alone. Today, apparel employment in the entire anthracite region totals about 3,000.[20] Similar trends have been seen in other manufacturing areas which in the 1960s lay claim to half of the labor force.

Today anthracite mining is less than a shadow of its former stature. About four hundred individuals work in some segment of the industry. Tonnage has declined from 2.5 million in the early 1990s to slightly more than one million in 1997, virtually none of it coming from the Wyoming and Lackawanna Valleys. Facing the loss of markets, anthracite producers continue to "downsize." In one example, the Jeddo Coal Company, a division of the Wilkes-Barre based Pagnotti Enterprises, saw its production decline from over 300,000 tons in the early 1990s to less than 40,000 tons today. Likewise, over the last ten years the firm's work force has declined from three hundred to sixty.[21]

Within the service economy, retailing has experienced strong growth as shopping malls and discount stores have proliferated. Banks, insurance companies, hotels, food processing firms, restaurants, building suppliers, and government offices have also expanded. Because the Scranton/Wilkes-Barre metro area ranks in the top ten in the percentage of population over sixty-five, health care and nursing home providers have become a significant part of the economy. So, too, has higher education as Luzerne and Lackawanna Counties are home to nearly a dozen colleges. In recent times travel and tourism has sprouted, in part because of the nearby Pocono Mountains (famous for vacation resorts and outdoor recreation); Scranton's Montage Mountain ski resort; the Scranton/Wilkes-Barre Red Barons and their minor league baseball stadium near Montage; Scranton's Steamtown National Historic Site; and the Lackawanna Coal Mine Tour and the Pennsylvania Historical and Museum Commission's Anthracite Heritage Museum, both located at McDade Park in Scranton. Indeed, both Lackawanna and Luzerne Counties have opened Offices of Tourism which now cooperate through the Northeast Territory Visitors Bureau. Soon Wilkes-Barre will build a sports arena complex and, by 2003, Scranton will boast passenger train service to New York City. These service sector ventures now account for the largest part of the economy and

TABLE 4. PERCENTAGE OF UNEMPLOYMENT IN THE NORTHERN ANTHRACITE FIELD, 1974-1998

Year	Luzerne County[22]	Lackawanna County[23]
1974	10.7	10.3
1976	11.3	11.0
1978	9.1	8.0
1980	11.4	9.2
1982	12.3	10.9
1984	10.3	9.1
1986	6.2	6.7
1988	6.4	4.8
1990	7.1	7.0
1992	10.3	8.9
1994	8.4	7.5
1996	7.0	6.9
1998	7.0	6.2

Fig. 52. Steamtown National Historic Site. (Courtesy of Ken Ganz, National Park Service, U.S. Department of the Interior)

have helped to keep the unemployment rate hovering near seven percent, which is still high compared to the national average of less than five percent (see table 4).

Recent economic trends have been characterized by continued "public-private partnerships." In the Wyoming Valley, the Greater Wilkes-Barre Chamber of Commerce has joined with the County Office of Community Development in fostering economic development. Similar public-private development efforts have been undertaken in Lackawanna County and have led to projects such as the Steamtown Mall in center city Scranton, formally opened in 1994 by Lackawanna County native and former governor Robert P. Casey, a determined advocate for the region's economic rebirth. Moreover, regional economic development has long been the concern of the non-profit Economic Development Council of Northeastern Pennsylvania, a pioneer in public-private cooperation.

Of course, service sector jobs do not generally pay as much, nor do they provide as many benefits as those in mining and manufacturing. Therefore, relatively low unemployment figures can be deceiving. Indeed, the average real wages of working people have declined locally and nationwide in recent decades, largely because of the growing service sector. Moreover, business locational decisions have been very uneven geographically, resulting in the growth of suburban shopping centers at the expense of places like downtown Wilkes-Barre which now has numerous vacant retail structures.

Today, leaders from communities throughout the anthracite region seek to build a new community image. They hope to make the "gritty coal town" perception an artifact of the past, buoyed by a diversified economy. To usher in the 1997 New Year, for example, community leaders in Wilkes-Barre applied a familiar tradition to send a signal that its image has changed. As the clock neared midnight on December 31, 1996, a chunk of anthracite coal descended from an erected platform much like the "big ball" in New York's Times Square. At precisely 12 o'clock, the coal was transformed into an illuminated diamond. Thousands of spectators packed Public Square in downtown Wilkes-Barre to view the event. The community was symbolically saying goodbye to its anthracite past and greeting its bright and hopeful future. (see fig. 53)

WHAT IF THE KNOX MINE DISASTER HAD NOT OCCURRED?

One question has persisted to the present: how long would anthracite mining have continued if the Knox Coal Company had not recklessly mined under the Susquehanna River? While the industry was certainly in decline, undoubtedly some deep mining would have remained for several years. Right up to the day of the disaster, large coal corporations were leasing mines to smaller companies, seasoned mineworkers were extracting coal, collieries were processing the product, and railroads and trucks were transporting it to consumers. Anthracite still had several markets and commanded a considerable share of the local economy. As mentioned in chapter 4, deep mining did continue until the early 1970s at the southern and northern parts of the field. When the national energy crises hit in the early and late 1970s, a more fully engaged anthracite industry surely would have been given a major boost.

However, even with extensive—and very profitable—strip mining operations, the anthracite markets could not be maintained yet alone expanded. In addition, it seems likely that the corruption that had gripped the industry, coupled with insufficient technological investments and the

Fig. 53. Revelers celebrate a lump of coal's transformation to a diamond, Public Square, Wilkes-Barre, New Year's Eve, 1996. (Courtesy of *Times Leader*)

gains by oil and natural gas, would have taken their toll in the long run. Therefore, it seems highly improbable that mining would have remained prominent.

Certainly had the disaster not occurred, twelve people would have been spared. The subsequent pain experienced by the community and the victims' families would have been avoided. And, the nationally recognized embarrassment of the ensuing accusations, indictments, trials, and convictions would have been missed. In the longer process of economic decline from anthracite, the Knox Mine Disaster revealed the shortcomings of the area's reliance upon a single industry and it confirmed the need for longer term planning for economic diversification. Had these efforts not been undertaken the regional economy would undoubtedly look much bleaker today.

Postscript

The Knox tragedy remains an important part of the northern coal field's history and culture. Its legacies are many. At the state level, several changes in mine laws were directly attributed to the calamity, including

Fig. 54. Knox Mine Disaster memorial at St. Joseph's Church, Port Griffith, Pa. (Photo by Robert P. Wolensky)

greater restrictions on mining under waterways. Within the community, the legacy includes vivid recollections of twelve deceased miners and their grieving families; crowds of spectators gathered along the banks of the river; wet and exhausted miners climbing out of the ground as family members welcomed each with tears of relief; gondolas and mine cars heaved into the Knox whirlpool and swallowed like bath toys; strong community support for widows and their dependents; and unyielding indignation over the human causes—greed, risk-taking, and corruption that had infected the Knox Coal Company and much of the anthracite industry. One victim's daughter captured the relentless grievance in a 1992 interview when she lamented, "I feel that they should call it the Knox Mine Murders."[24]

Economic development emerged as another legacy of the disaster. Suddenly, the reality of a future without anthracite had been thrust upon the community. Industrial diversification efforts were now more important than ever.

The memory of the Knox Mine Disaster has stayed with the citizens of the northern field. It has been commemorated by a monument in front of St. Joseph's Catholic Church in Port Griffith listing the names of the deceased, and by an annual mass at the church that has included participation by survivors as well as the victims' families.

Local newspapers carry anniversary stories typically accompanied by first person interviews and now legendary photographs. In 1996, Luzerne County Community College devoted the entire afternoon portion of its annual History of Northeastern Pennsylvania Conference to a panel discussion on the Knox catastrophe. The Pennsylvania Historical and Museum Commission, in conjunction with local sponsors, has placed a permanent state historical marker near the disaster site in commemoration of its fortieth anniversary (see back cover). The Anthracite Heritage Museum in Scranton opened a fortieth anniversary exhibit on the disaster which will run through 1999.

Coal mine disasters, like most other man-made calamities, rarely have happy endings. No doubt the northern anthracite region's struggle to understand and reconcile the causes and consequences of the epoch-ending Knox Mine Disaster will continue into the future.

NOTES TO CHAPTER FIVE

1. George M. Leader, taped interview, May 30, 1995, WVOHP, tape 1, side 1. Leader served as governor from 1955 to 1959.

2. Donald Miller and Richard Sharpless, *The Kingdom of Coal*, 1985.

3. "LIFE Volunteers Open Fund Drive in County: $1,500,000 Sought," *Scranton Times*, April 21, 1959, 3, 18; "LIFE Fund $119,824 Bigger," *Scranton Times*, May 1, 1959, 1, 20.

4. Jacob Kaufman and Helmut Golatz, *A Study of Areas of Chronic Unemployment in Pennsylvania* (University Park: Pennsylvania State University Press, 1959).

5. Jacob Kaufman and Helmut Golatz, *A Study of Areas of Chronic Unemployment in Pennsylvania, 1959*. See also Pennsylvania Department of Labor and Industry, Bureau of Employment Security, *Labor Market Letter, Wilkes-Barre/Hazleton Area*, November, 1954, vol. IX, no. 11; December 1956, vol. XI, no. 12; December, 1958, vol. XIII, no. 12; December, 1960. vol. XV, no. 12; December 1962, vol. XVII, no. 12; December, 1964, vol. XIX, no. 12; November, 1966, vol. XXI , no. 12; November, 1968, vol. XXIII, no. 11; December, 1970, vol. XXV, no. 12; December, 1972, vol. XXVI, no. 12; December, 1974, vol. XXIX, no. 12; December, 1976, vol. XXXI, no. 12; December, 1978, vol. XXXIII, no. 12; December, 1980, vol. XXXV, no. 12; December, 1982, vol. IX, no. 6; and December, 1984, vol. XI, no. 4. Harrisburg, Pennsylvania: Department of Labor and Industry, Bureau of Employment Security, *Labor Market Letter, Northeastern Pennsylvania*, December, 1986, vol. II, no. 4; December, 1990, vol. VI, no. 4. Harrisburg, Pennsylvania, Department of Labor and Industry, Bureau of Research and Statistics, *Current Labor Market Information Bulletin, Wilkes-Barre/Scranton*. August 7, 1992; January 7, 1994. Data for 1996 and 1998 was provided by the Department of Labor and Industry, Bureau of Research and Statistics.

The authors' intent was to report data as close to year-end as available. Unless otherwise cited in this note, all data reported from the Department of Labor and Industry is for the month of December.

6. Pennsylvania Department of Labor and Industry, Bureau of Employment Security, Labor Market Letter, *Scranton and Lackawanna County*, March 1953, vol. VII, no. 3; November 1954, vol. IX, no. 11; November 1956, vol. XI, no. 11; November 1958, vol. XIII, no. 11; December 1960,

vol. XV, no. 12; December 1962, vol. XVII, no. 12; December 1964, vol. XIX, no. 12; December 1966, vol. XXI, no. 12; November 1968, vol. XXIII, no. 11; December 1970, vol. XXV, no. 12; December 1972, vol. XXVII, no. 12; December 1974, vol. XXVIII, no. 12; December 1976, vol. XXIX, no. 12; December 1978, vol. XXXI, no. 12; December 1980, vol. XXXIII, no. 12; December 1982, vol. XXXV, no. 12. Also, Pennsylvania Department of Labor and Industry, *Labor Force Employment, Unemployment and Unemployment Rate for the United States, Pennsylvania, and Major Areas of Pennsylvania.* October 1985; December 1986; December 1988; April 1991; February 1992; December 1994, State Library of Pennsylvania, Government Documents Section. Data for 1996 and 1998 was provided by the Department of Labor and Industry, Bureau of Research and Statistics.

The authors' intent was to report data as close to year-end as available. Unless otherwise cited in this note, all data reported from the Department of Labor and Industry is for the month of December.

7. On the Wyoming Valley's industrial development see Richard Cronin, taped interview, May 24, 1983, WVOHP; Harold Landau, "Industrial Development in the Wilkes-Barre Area," master's thesis, University of Scranton, 1967; Sheldon Spear, *Wyoming Valley History Revisited* (Shavertown, Pa. : JEMAGS & Co. Publishing, 1994), especially pages 224-236. For an overview of the valley's economic transition from demise of anthracite to the modern service economy see Kenneth C. Wolensky, "Diamonds and Coal," *Now and Then—The Appalachian Magazine*, 14 (Winter 1997), 20-24.

8. Commonwealth of Pennsylvania, Office of Budget and Administration, *The Pennsylvania Industrial Development Authority: An Assessment*, Harrisburg, 1979.

9. Commonwealth of Pennsylvania, Office of Budget and Administration, *The Pennsylvania Industrial Development Authority: An Assessment* Harrisburg, 1979. Also, for a detailed analysis and critique of the impact and results of PIDA funding to the anthracite region see Thomas Dublin, "Attracting Business to the Anthracite Region, 1940-1970: Promises and Performance," paper presented to the Hagley Research Seminar, Wilmington, Delaware, May 8, 1997.

10. George M. Leader, taped interview, May 30, 1995, WVOHP, tape 1, side 1.

11. George M. Leader, taped interview, May 30, 1995, WVOHP, tape 1, side 2.

12. William C. Kashatus III, "'Dapper Dan' Flood, Pennsylvania's Legendary Congressman," *Pennsylvania Heritage* 23 (Summer 1995), 4-11.

13. For further information on Congressman Flood, his influence over the federal budget, and stature in the local community and in Washington see: William Kashatus III, "'Dapper Dan' Flood, Pennsylvania's Legendary Congressman," 1995; Miller and Sharpless, *The Kingdom of Coal*, especially chapter 9 and Epilogue; and George Crille, "The Best Congressman," *Harper's Magazine*, (January 1975), 11-16.

14. Like Congressman Flood (See chapter 1, note 22), McDade also became mired in scandal; nevertheless, his constituents have appreciated his work on behalf of the area, and they have remained loyal. After serving thirty-six years in Congress, McDade retired in January, 1999.

15. Commonwealth of Pennsylvania, Office of Budget and Administration, *The Pennsylvania Industrial Development Authority: An Assessment*, Harrisburg, 1979; Thomas Dublin, "Attracting Business to the Anthracite Region, 1940-1970: Promises and Performance," paper presented to the Hagley Research Seminar, Wilmington, Delaware, May 8, 1997.

16. See note 4.

17. See note 5.

18. On the Wyoming Valley's recovery from Tropical Storm Agnes see Robert P. Wolensky, *Better Than Ever: The Flood Recovery Task Force and the 1972 Agnes Disaster* (Stevens Point: University of Wisconsin Stevens Point Foundation Press, 1993).

19. Rachel L. Swarns, *The Wall Street Journal*, July 14, 1987, 35.

20. For further discussion of the garment industry in the Wyoming Valley, the role of the International Ladies' Garment Workers' Union, and the industry's decline, see Robert P. Wolensky and Kenneth C. Wolensky, "Min Matheson and the ILGWU in the Northern Anthracite Region," *Pennsylvania History (Special Issue on Oral History)* 60 (1993), 455-474; Kenneth C. Wolensky and Robert P. Wolensky, "Building the ILGWU in Pennsylvania's Anthracite Mining Towns: The Leadership of Min Matheson, 1944-1963," *Sociological Imagination*, 31 (1994), 83-100. Also see Kenneth C. Wolensky, "We Are All Equal: Adult Education and the Transformation of Pennsylvania's Wyoming Valley District of the

ILGWU, 1944-1963," unpublished doctoral dissertation: Pennsylvania State University, 1996.

21. Louis Pagnotti III, taped interview, January 11, 1998, WVOHP.

22. See note 4.

23. See note 5.

24. Robert P. Wolensky and Kenneth C. Wolensky, "Disaster—Or Murder?—In The Mines," *Pennsylvania Heritage* 24 (Spring 1998), 4-11.

APPENDIX I
ORAL HISTORY INTERVIEWS

Adonzio, Charles (03-14-96), family anthracite mining business.

Amos, Hubert (01-12-95), mine foreman, Harry E. Coal Co.

Baloga, Donald (01-12-95), son of Knox Mine Disaster victim (John Baloga).

Bednark, Margaret Maloney (06-10-98), daughter of murdered UAMP leader Thomas Maloney.

Brominski, Judge Bernard, (05-31-91; 10-10-95), presided at a Knox Mine Disaster case in Luzerne County.

Brozena, Chester (12-03-88), anthracite mineworker, Glen Alden Coal Company, who participated in the United Anthracite Miners' (UAMP) strikes in the 1930s.

Borosky, Edward (08-04-89), survivor of the Knox Mine Disaster.

Burns, Francis (01-21-92), son of Knox Mine Disaster victim (Francis Burns).

Burns, Thomas (08-08-89), survivor of the Knox Mine Disaster.

Butler, William (07-25-94), coal operator in the Hazleton area.

Butsavage, Leo (08-07-89), survivor of the Knox Mine Disaster.

Calvey, Audrey (06-17-92), daughter of Knox Mine Disaster victim

(John Baloga).

Cappellini, Marie (12-14-88; 12-09-88), wife of labor leader Renaldo Cappellini.

Carey, Melbourne (01-14-88), anthracite mineworker.

Carmon, Raymond (05-31-83), economic development specialist.

Casterline, Lewis (12-12-88; 05-25-90), mine superintendent, Pennsylvania Coal Company, UAMP member, and private insurance inspector.

Chamberlain, Alex (07-11-95), family anthracite mining business and mining engineer.

Cherkowski, Edward (08-04-89), anthracite mineworker, Scranton area companies.

Connor, Thomas (01-22-98), real estate officer, Pagnotti Enterprises.

Conyngham, Guthrie (08-01-89), businessman and family coal interests.

Conyngham, Jack (John) (08-04-94), businessman and family coal interests.

Conyngham, William (08-02-91), businessman and family coal interests.

Costa, Joe (Costanzio Lopez) (08-14-85), mining contractor in the 1930s, and employee of various coal companies.

Danna, Frank (05-18-94), Pagnotti company superintendent who assisted with Knox disaster recovery.

D'Angelo, Anthony (12-17-88; 07-28-89; 02-17-94), anthracite history and culture.

DeGillio, William (07-03-90), defendants' attorney in two Knox Mine Disaster trials.

Dorish, John (07-31-88), anthracite history and culture.

Dubee, Michael (01-11-93), anthracite mineworker, Lehigh Valley, Glen Alden, and Delaware and Hudson coal companies.

Dukes, John (01-22-94), office employee, payroll, Glen Alden Coal Company.

Dunn, Chester (08-04-89), survivor of the Knox Mine Disaster.

Embleton, Arnold (10-07-96), Lehigh Valley Railroad conductor.

Everett, William (01-13-95), anthracite history and culture, Glen Alden Coal Company.

Featherman, Opal (06-10-97), wife of Knox Mine Disaster victim

(Charles Featherman).

Ferrare, Frank (06-21-92), son-in-law of Knox Mine Disaster victim (Sam Altieri).

Ferrare, Anne (06-21-92), daughter of Knox Mine Disaster victim (Sam Altieri).

Ferraro, Michael (10-29-88), coal operator, Swoyersville, PA.

Flood, Rep. Daniel (07-05-90), congressional representative, 11th District of PA.

Francik, Joseph (06-26-90), survivor of the Knox Mine Disaster.

Gadomski, John (12-22-88; 06-28-90), survivor of the Knox Mine Disaster.

Gatti, Gerald (07-14-95), Gatti Engineering, Inc. dug the River Slope entrance.

Gizenski, Ida (07-10-97), wife of Knox Mine Disaster victim (Joseph Gizenski).

Gizenski, Joseph (07-10-97), son of Knox Mine Disaster victim (Joseph Gizenski).

Gizenski, Al (7-10-97), son of Knox Mine Disaster victim (Joseph Gizenski).

Graham, William (07-11-95), chief executive of the Penn Anthracite Collieries Company, Scranton.

Gushanas, George (01-13-94), Glen Alden Coal Company superintendent who assisted with Knox disaster recovery.

Guzo, John (06-16-92), anthracite mineworker and foreman, Harry E. Coal Company.

Hague, William (08-07-89), survivor of the Knox Mine Disaster.

Handley, Frank (12-10-88; 06-27-90), survivor of the Knox Mine Disaster and foreman at the Knox Coal Company and other coal companies.

Harenza, Jacob (07-30-92), brother-in-law of Knox Mine Disaster victim (John Baloga), and carpenter for the Knox Coal Company.

Hastie, William (07-31-89; 06-28-90; 07-02-90; 05-23-91; 03-13-96), anthracite mineworker, Knox Coal Company.

Kanaar, Al (10-28-88), Knox Coal Company assistant foreman, rock contractor.

Kopcza, Joseph (12-21-88), survivor of the Knox Mine Disaster.

Kupstas, Agnes (12-04-88), personal secretary to Msgr. John J. Curran, the "labor priest."

Lazar, Peter (06-26-90), anthracite mineworker, Harry E. Coal Company.

Leader, George (May 30, 1995), former governor of Pennsylvania, 1955-59.

Lucas, Michael (06-25-90), survivor of the Knox Mine Disaster.

Matheson, Min L. (11-30-82; 12-05-88; 06-28-90), labor leader for the ILGWU in the Wyoming Valley.

Mazur, George (12-20-88), survivor of the Knox Mine Disaster.

Meletsky, Ambrose (12-05-82; 12-16-82), anthracite region activist.

Milner, Stewart (08-03-92), vice president, Pennsylvania Coal Co.

Moss, Mary Mitchell (07-31-91), granddaughter of UMWA president John Mitchell.

Noterman, Joseph (08-18-83), anthracite region activist.

Obsitos, Michael (12-08-88), survivor of the Knox Mine Disaster.

Ogin, Anita (07-29-92), daughter of Knox Mine Disaster victim (Eugene Ostrowski).

Ostrowski, Donna (07-29-92), daughter of Knox Mine Disaster victim (Eugene Ostrowski).

Ostrowski, Eugene (07-29-92), son of Knox Mine Disaster victim (Eugene Ostrowski).

Orlowski, Charles (06-09-98), anthracite mineworker.

Panzitta, David (03-14-96), family anthracite mining business.

Pagnotti, Louis III (01-11-98), still active family anthracite mining business.

Parente, Yolande Altieri (06-21-92), Daughter of Mine Disaster Victim (Sam Altieri).

Piasecki, Anna (06-04-91), wife of UMWA Local 8005 union official.

Pincoski, Helen ("Mike") (12-07-88), anthracite history and culture.

Randall, David (06-07-98), anthracite mining family and coal operator in the Southern Field.

Remus, Anthony (12-28-88), survivor of the Knox Mine Disaster.

Roman, Stanley (12-22-88), survivor of the Knox Mine Disaster.

Silverblatt, Arthur (07-24-89), assistant district attorney, Luzerne County, 1959.

Siracuse, Angelo (07-19-89), anthracite history and culture.

Siracuse, Jennie (08-14-85), anthracite history and culture.

Slipetz, Michael (05-30-91), anthracite history and culture.

Slusser, Ronald (12-29-88;12-30-88), anthracite history and culture.

Sporher, George (07-03-90), anthracite history and culture.

Stark, Leah (09-14-90), wife of Knox Mine Disaster victim (William Sinclair).

Stefanides, Daniel (08-01-94), son of Knox Mine Disaster victim (Daniel Stefanides).

Stefanides, Joseph (09-15-90), Knox Coal Company employee and brother of Knox Mine Disaster victim (Daniel Stefanides).

Stefanides, Stephanie (09-15-90), wife of Knox Mine Disaster victim (Daniel Stefanides).

Stella, Joe (Pacifico) (11-01-88), Pennsylvania Coal Company inspector at the River Slope mine.

Suchocki, Alfrede (05-20-94), member, Knox Mine Disaster Memorial Committee.

Swan, Preston "Pep" (01-15-89), anthracite mineworker.

Tedesco, James (10-08-96), anthracite coal operator, Pagnotti Enterprises.

Teller, Stephen (06-27-90; 01-05-95), district attorney, Luzerne County, 1959; prosecuted Knox Mine Disaster cases.

Thomas, Robert (05-28-96), son of Myron Thomas, Knox Mine Disaster survivor.

Thomas, Tom (01-19-98), Office employee, maps, Glen Alden Coal Company.

Throne, Hank (09-11-97), survivor of the Shepton Mine Disaster in 1963.

Urban, Bonnie (07-21-89), anthracite mineworker, Harry E. Coal Company.

Volpe, Charles (1-22-98), anthracite mineworker, various companies.

Waitkevich, Anthony (08-02-89), survivor of the Knox Mine Disaster.

The findings and conclusions expressed in this book are not necessarily those of the people who contributed oral history interviews. Not all of the interviews listed were cited in the book, but each person provided important insights on twentieth century anthracite history and culture.

These oral history interviews are part of the authors' three hundred-person Wyoming Valley Oral History Project which is housed at the University of Wisconsin-Stevens Point.

APPENDIX II
GLOSSARY OF ANTHRACITE MINING TERMS

anticline. A fold in a rock and coal seam in which the opposite sides of the fold dip away from each other like the two legs of the letter A. The effect also resembles a saddle.

barrier pillars. The solid blocks of coal left in the ground to separate adjacent collieries, particularly to prevent a flood or other disaster in one from spreading to the other. *See also* **robbing the barriers**.

borehole. A small opening through rock and coal strata made from the surface by a drilling machine. A borehole is used determine the nature of the coal seams or to provide an opening for pipes or electric cables that can be passed into the mines.

breaker. The tall building into which coal travels directly from the mine and wherein the coal is broken, sized, and washed before shipment to market.

Buried Valley of the Susquehanna. Between the river bottom and the top rock of the anthracite coal beds in the Wyoming Valley lies a heavy layer of sand, clay, and gravel left by the erosive action of an ancient glacier. These deposits, called the Buried Valley of the Susquehanna, extend to a depth of up to 320 feet and come precariously close to the coal seams in some areas. These water-bearing, quicksand-like sediments follow the Susquehanna River for fifteen miles through the Wyoming Valley and present a potential hazard for mining operations near the waterway. Companies had to pay special attention to the thickness of roof cover or top rock separating a mine chamber from the buried valley. Should the roof get too thin, the buried valley could break into the mines, as it did with the Knox Mine Disaster, causing a flood of major proportions. *See also* **ceiling**, **roof**, and **top rock**.

cage. The elevator platform within a mine shaft used to transport mineworkers, equipment, as well as loaded and empty mine cars into and out of the mine. *See also* **shaft**.

ceiling. The roof composed of rock lying above a mine chamber. *See also* **roof**.

chamber. A miner's work place, also referred to as a room, breast, or place. *See also* **room** and **place**.

colliery. The numerous surface buildings along with the underground mines that constitute a complete mining operation.

contract-leasing system. Beginning in the 1920s, the large anthracite coal companies that controlled the mineral rights in the northern field began (sub)contracting sections of their mines to individual entrepreneurs who hired crews of up to twenty workers. The entrepreneurs often violated the collectively bargained agreement with the United Mine Workers union and played fast-and-loose with safety procedures. However, they boosted production and lowered costs in the short run. During the 1930s the large firms—with the Pennsylvania Coal Company leading the way—went beyond contracting and began leasing whole mines and even entire collieries to independents who were now incorporated. Many, though by no means all, of the Pennsylvania Coal Company's leases were issued to individuals with alleged organized crime affiliations. The Knox Coal Company, which leased from the Pennsylvania Coal Company, was owned by some individuals with alleged ties to organized crime. Each of the region's five large companies—The Pennsylvania, Lehigh Valley, Hudson, Glen Alden, and Susquehanna—participated in the contract-leasing system. *See also* **lessor companies** and **lessee companies**.

contractors. Independent entrepreneurs who secured agreements or (sub)contracts from large coal companies to take coal in certain sections of a mine or to perform specific tasks such as digging a rock tunnel. Contracting first became a major issue at the Pennsylvania Coal Company and elsewhere in the northern field during the early 1920s when workers protested the practice as a threat to their job security and wage rates.

crosscut. A passageway driven at right angles to connect with a working chamber. Used to enhance ventilation.

dead work. Exploratory or preparatory work such as removing rock, cleaning roof falls, and setting props, during which time little or no coal is taken.

first mining. The removal of virgin coal from a seam using the room-and-pillar method of mining. *See also* **seam** and **room and pillar method**.

forecast. A written request by a lessee company to its lessor seeking permission to mine coal in an area not covered by the original lease. *See also* **lessee companies** and **lessor companies**.

gangway. The main underground rail haulage road in an anthracite mine.

lessee companies. Small to mid-sized anthracite coal companies that did not own or control the mineral rights, so they leased coal properties from lessor companies who did control the mineral rights. The leases were often issued for second minings. *See also* **lessor companies, contract-leasing system**, and **second mining**.

lessor companies. Large anthracite coal companies that owned or had control of mineral rights. They leased coal properties to lessee companies because the lessees disciplined the workers, corrupted the union, lowered costs, boosted production, and otherwise engaged in mining practices that the more established and prestigious lessors could not. *See also* **lessee companies** and **contract-leasing system**.

make water. The water that continuously seeps into the mines.

Northern Anthracite Field. The anthracite coal deposits in the Luzerne County/Wilkes-Barre and Lackawanna County/Scranton areas.

pillar. A solid block of coal left between mined-out chambers in the room-and pillar method of mining. *See also* **room and pillar method** and **robbing the pillars**.

place. The miner's working chamber. *See also* **room** and **chamber**.

prop. A wooden log cut and set to support the roof within a mine chamber. *See also* **chamber** and **roof**.

robbing the barriers. The usually illegal removal of the barrier pillars of solid coal left in the ground to separate the workings of adjoining collieries. *See also* **barrier pillars**.

robbing the pillars. The removal of coal pillars, usually a legal undertaking, was conducted without regard for surface subsidence, which often occurred. *See also* **pillars** and **room-and-pillar-method**.

roof. The "ceiling" or bottom layer of top rock within a underground working place. *See also* **ceiling** and **top rock**.

room. The miner's chamber or working place. *See also* **place** and **chamber**.

room-and-pillar method. A method of mining used in the northern field whereby solid blocks of coal were left between mined out chambers. The blocks, called pillars, were used to support the roof until the first mining was finished whereupon a second mining was undertaken to remove the pillars. *See also* **chamber** and **second mining**.

seam, coal. The strata of anthracite, also know as a coal bed.

second mining. The removal of the solid blocks, or pillars, of coal separating mined-out chambers. *See also* **robbing the pillars** and **room-and-pillar method**.

shaft. A vertical opening dug into a seam of coal into which two parallel elevator cages were built. Used to transport mineworkers, mine cars, and equipment. Also used as the exit through which air flowed into the mines and water was pumped out. *See also* **cage**.

slope. An inclined opening into a seam of coal, laid with rails, and used to transport mineworkers, mine cars, and equipment into and out of the mine.

sweetheart contract. A corrupt labor-management agreement whereby certain terms of the collectively bargained contract are violated. In the typical case, management offers and a union official accepts special compensation or other rewards in return for lax or non-enforcement of the agreement. It is, in effect, a conspiracy to circumvent a legally binding workplace document.

top rock. The strata of rock lying above an anthracite seam of coal. *See also* **ceiling** and **roof**.

yardage. A system whereby mineworkers were paid by the lineal yard. Often used do determine payment for dead work associated with removing rock. *See also* **dead work.**

This glossary draws upon mining terms contained in the Hudson Coal Company's *The Story of Anthracite* published by the company in 1932; and *A Dictionary of Mining, Mineral, and Related Terms*, compiled and edited by Paul W. Thrush and the staff of the U.S. Bureau of Mines, Department of the Interior, 1968.

INDEX

Agnes Recovery Act of 1972, 134
Alaimo, Dominick, 88, 89, 108, 115
Altieri family, 32
Altieri, Samuel, 28, 35
Anthracite Heritage Museum, 136, 142
anthracite industry
    Anthracite Task Force, 132
    Anthracite Unemployed Worker, 126
    cartoons, Knox Mine Disaster, 67, 68
    contract-leasing system, xvi, 102-10
    corruption, xvi, 109-10.
    deaths in anthracite mining, 84
    disasters in anthracite mining, 7, 37-39, 69, 72,
    "dogholes," 126
    geology, 72-74, 75, 77, 81
    mining laws and regulations, 70-71, 72-73, 87, 113-14
    padrone system, 102
    production statistics, 69, 125
    safety, 108
anthracite region, 5-6
    economic decline, 126-33, 139
    outmigration, 126
    subsidence (mine caving), 69
anticline, 74, 75, 77
Apalachin (N.Y.) crime meeting, 88
Argo, Anthony, 89, 90
Army Corps of Engineers, 46
Ash, S.H., 72
Aston, Albert, 87
Avoca (Wilkes-Barre/Scranton) Airport, 131
Avon Coal Company, 94
Baloga, Donald, 42
Baloga family, 29
Baloga, John, 29, 35
Bernardi Coal Company, 108
Blue Coal Corporation, 97, 111
Bohn, Fred, 9, 10
bombings (Good Friday), 105
bootleg mining, 79, 126
Borosky, Ed, 26
Boyar, Benjamin, 29, 35
Boyle, W.A. "Tony," 98
Brennan, Thomas J., 88
Brominski, Judge Bernard J., 93, 96
Bufalino, Russell, 114
Buried Valley of the Susquehanna, 72-74, 81
Burke, Thomas, 92
Burns, Francis, 29, 35
Burns, Francis Jr., 42
Calvey, Audrey, 42, 43
CAN DO (Hazleton), 126, 131
Capone Coal Company, 108
Cappellini, Rinaldo, 102
Cardoni, Frank, 118
Carter, Pres. Jimmy, 132
Casey, Gov. Robert P., 138
Casper, Charles, 94
Cecconi, Fred, 12
Chamber of Commerce, Greater Wilkes-Barre
    The Committee of 100, 126
    Crestwood Industrial Park, 128, 132

economic development, 138
   Industrial Development Fund, 126
   New Industries Committee, 126
Cigarski, Steven, 24
cofferdam, 55-56
Como, Perry, 70
Conlon Coal Company, 66
Connolly, Daniel, 14
corruption, xvi, 109-10
Costa, Joe, 109
costs of Knox Mine Disaster, 63-64
court cases, 89-97, 101
Coxton Yard, 46, 47
Crestwood Industrial Park, 128, 132
Curran, Msgr. John J., 104
Danbury (Ct.) Federal Prison, 93, 94
Danna, Frank, 46, 57, 58
DeAngelo, Tony, 120
Degillio, William, 95
Dialeda, George, 91, 97
Dialeda, Norman, 97
disasters, 7, 37-39, 69, 72
Domoracki, Frank, 9, 19
Dorrance Colliery, 67
Dougherty, Robert L., 3, 74, 85, 89, 90, 91, 92, 93, 94
Dougherty, Robert L. Jr., 92
Drustrup, Captain Norman J., 45
Dublin, Thomas, 120-21, 144
Dunmore, 77
Dunn, Chester, 40
Durkin, James, 111
Eagle air shaft, 14, 20, 21, 25

Earth Conservancy, Inc., 111
Eberhard Faber, 128
Economic Development Council of Northeastern Pennsylvania, 138
Eisenhower, Pres. Dwight D., 53
Embleton, Arnold, 46
Enterprise Colliery, 66
Ewen Colliery, 8, 9, 103
Exeter National Bank, 95
Fabrizio, Louis, 3, 85, 89, 90, 91, 92, 93, 94, 108
Featherman, Charles, 9, 35
Featherman, Opal, 43
Federal-State Mine Drainage Program, 1955, 54
Ferrare, Frank, 42
Ferrare, Anne, 42
Fine, Gov. John S., 122
Flood, Rep. Daniel J., 14, 130-32
Flood-Douglas Area Redevelopment Bill, 132
Fortney, Gerald, 22, 23
Foster-Wheeler Corporation, 128
Franklin Colliery, 57
Fries, Ralph, 74, 75. 87, 90
Fugmann, Michael, 121
Gadomski, John, 21, 108
garment industry, 126, 135-36
Gatti Engineering Co., 8
Gaul, Gil, 111
Gizenski, Al, 43
Gizenski, Ida, 43
Gizenski, Joseph Jr., 43
Gizenski, Joseph Sr., 9, 10, 35
Glen Alden Coal Company, 4, 48, 54, 67, 104, 105, 111

glossary of anthracite mining terms, 153-57
Goddard, Maurice, 56
gondolas (railroad cars), 40, 46, 48, 57, 58, 61-62
Good Friday bombings, 105
government, local, 69
Graham, William, 106, 120
Groves, Robert, 11, 13, 42, 74, 87, 91, 93, 95
Gushanas, George, 48, 109
Hague, William, 40
Handley, Frank, 2, 11-12, 13, 22, 56, 59, 87
Harenza, Jake, 43
Harvan, George, 48
Hastie, William, 15, 16, 17, 40
Hazleton, 126, 130, 131
Henry Colliery, 65
Hoffa, James R., 97, 111
Huber Colliery, 109
Hudson Coal Company, 4, 54, 67, 104
indictments in Knox Mine Disaster, 88-89
inspectors, mine, 109
International Ladies' Garment Workers' Union (ILGWU), 70, 80-81, 129
Interstate 81 (I-81), 131
Jamieson, James, 40
Jaspin, Elliot, 111
Jeddo Coal Company, 136
Jeffrey, James, 87
Kanaar, Al, 83
Kanjorski, Rep. Paul, 111
Kaveliskie, Dominick, 35

Kennedy, Pres. John F., 130
Kennedy, Joseph T., 54, 56
Kennedy, Thomas, 97, 115
King Fifth Wheel, 128
Knox Coal Company, 2, 28, 74, 78, 89, 91, 100, 107, 108, 119, 139, 142
Knox Mine Disaster monument, 141, 142
Knox Mine Disaster state historical marker, 142, back cover
Kopriver, Rep. Frank, 83
Kulikowich, Alex, 2
Kupcza, Joseph, 40
labor-management relations
 conflict, 104-5, 106
 United Anthracite Miners of Pennsylvania, 104
Lackawanna Basin, xv, 5, 62-63
Lackawanna Coal Mine tour, 136
Lackawanna County
 employment in anthracite mining, 1950-1998, 125, 127
 Northeast Territory Visitors Bureau, 136
 population, 80
 tourism promotion, 136
 unemployment, 1950-1972, 133; 1974-1998, 137
Lackawanna Industrial Fund Enterprise (LIFE), 126
Lackawanna River, 72
Langan, Fr. Edmund, 26, 27
Lawrence, Gov. David L., 28, 53, 56, 68
Leader, Gov. George M., 125, 128-30

Lehigh Valley Coal Company, 4, 65, 67, 104, 110
Lehigh Valley Railroad, 2, 46, 47, 50
Lewis, John L., 97, 104, 115
Lippi, August J., 85, 86, 89, 90, 91, 92, 95, 97, 98, 99, 100
Lippi, John, 98
Lucas, Mike, 2, 28, 29
Luzerne County
    employment in Anthracite Mining, 1950-1998, 126, 127
    Northeast Territory Visitors Bureau, 136
    Office of Community Development, 138
    population, 80, 126
    Redevelopment Authority, 134
    tourism promotion, 136
    unemployment, 1950-1972, 133; 1974-1998, 137
Luzerne County Community College, 142
Maguire, Arthur, 92
Maguire, B. Todd, 92
Maloney, Thomas, 104, 105
Martin, Joseph A., 86
Mazur, George, 11, 20, 21
McDade, Rep. Joseph, 132
McGuigan, Frank, 95
Milner, E. Stewart, 103, 107
mining laws and regulations, 70-71, 72-73, 87, 113-14
Moffat Coal Company, 111
Montage Mountain, 136

Mulhall, John, 95
National Labor Board, 104
Navy Construction Battalion, 46, 108
Nixon, Pres. Richard M., 134
Northeast Territory Visitors Bureau, 136
No. 1 Contracting Co., 55
No. 14 Colliery, 66
Obsitos, Michael, 40
off-course mining under the Susquehanna River, 76-77
Ogin, Anita, 43
organized crime, 88, 101, 107-8 114, 115
Orlowski, Frank, 26, 35
Orlowski, Theresa, 26
Osticco, James, 114
Ostrowski, Donna, 43
Ostrowski family, 32, 34, 36
Ostrowski, Eugene Jr., 43
Ostrowski, Eugene Sr., 2, 9, 21, 34
Pagnotti Enterprises, 54, 55, 57, 136
Pagnotti, Louis Sr., 110, 111
Pancotti, Amadeo, 14, 15-16, 19, 20
Parente, Yolanda, 42, 43
Payne Coal Company, 108
Peeley Coal Company, 94
Penn Anthracite Mining Company, 106
Pennsylvania Coal Company, 4, 8, 9, 66, 71, 74, 77, 102, 103, 104, 105, 107, 119
Pennsylvania, Commonwealth of

Crime Commission, 107, 122
Coal Mining Law of 1891, 66
Commission of State Mine
 Inspectors, 87
Department of Mines and
 Mineral Industries, 73
Historical and Museum
 Commission, 136, 142
Industrial Development
 Authority (PIDA), 128-30
Joint Committee to Investi-
 gate the Knox Mine
 Disaster, 85, 101
Piasecki, Charles, 89, 90
Pittston Vein, 8, 74, 105
Pocono Mountains, 136
Poluske, Joseph, 2
Port Griffith, 2, 70, 71, 112, 141, 142
Port Griffith Disaster Fund
 Committee, 70
Pulitzer Prize (*Pottsville Republi-
 can*), 111
Quinn, Vincent, 94
Radio Corporation of America
 (RCA), 132
Receski, William, 74, 75, 77, 87, 90, 91, 93, 95
Red Barons, Scranton/Wilkes-
 Barre, 136
Remus, Anthony, 40
Renner, Fritz, 74, 75, 87, 90
River Slope Mine, 2, 8, 18-19, 22, 27, 52, 54, 59, 60, 68, 69, 74, 76, 105
robbing the barrier, 66, 79
robbing the pillars, 69-70

Roberts, Ellis W., 80
Roosevelt, Pres. Theodore, 104
safety, 108
St. Cecilia's Church (Exeter), 85
St. Joseph's Catholic Church (Port
 Griffith), 141, 142, back cover
Saporito Coal Company, 107
Schooley Colliery, 8
Sciandra, John, 107, 108, 122
Sciandra, Angelo Joseph, 114
Sciandra, Josephine, 93, 94
Scranton
 Scranton Times, 98
 Steamtown Mall, 138
 Steamtown National Historic
  Site, 132, 136, 138
 University of Scranton, 132
sealing the breach, 55-59
Shapp, Gov. Milton J., 134
Shirey, Warren, 22
Sinclair, William, 12, 29, 35, 42
Sorber, Al, 48
Stark, Lea, 43
Statkewicz, Leonard, 118
Steamtown Mall, 138
Steamtown National Historic Site, 132, 136, 138
Stefanides, Daniel Jr., 31, 33, 43
Stefanides, Daniel Sr., 12, 29, 33, 35
Stefanides family, 31, 33
Stefanides, Joseph, 43
Stefanides, Stephanie, 43
Stella, Joe (Pacifico), 17-20, 21, 22, 41, 77
stop lines, 73, 75, 76
Susquehanna Colliery Company,

67, 72, 104, 105, 111
Susquehanna River, 1-2, 15, 51, 112, 134
Taft-Hartley labor law, 89, 91, 92, 93
Tedesco, James, 110
Teller, Stephen A., 90, 91, 94
Thomas, Lester, 98
Thomas, Myron, 11, 17-20, 22, 24, 25
Thomas, Robert, 41
Tropical Storm Agnes, 134-35
unemployment councils, 126
United Anthracite Miners of Pennsylvania, 104
United Mine Workers of America, xvi, 4, 84, 88, 98, 100, 101, 102, 107
United States government
    Appalachian Regional Commission, 130
    Area Redevelopment Administration, 130
    Bureau of Mines, 65, 87
    Coal Mine Health and Safety Act of 1969, 131
    Department of Commerce, 130
    Department of the Interior, 132, 138
    Federal Bureau of Investigation, 88
    Geological Survey, 52, 68
    Office of Anthracite, 132
    Special Group on Organized Crime, U.S. Justice Department, 88

U.S. Steel Corporation, 54
Tobyhanna Army Depot, 131, 132
University of Scranton, 132
Valmont Industrial Park, Hazleton, 130, 131
Volpe, Santo, 107, 122
Waitkevich, Anthony, 79
Wall Street Journal, 98
water
    de-watering the River Slope, 53-55, 110
    entering River Slope Mine, 52, 110
    pump installation, 54-55
    Water Hazards Commission, 72-73
    water problems in anthracite, 53-54
Wilkes-Barre
    Chamber of Commerce, 126, 128, 132, 138
    City Redevelopment Authority, 134
    Public Square, 134, 135, 139, 140
Williams, John, 1, 9, 10, 42
Wyoming Basin, xv, 5, 71, 74
Wyoming Valley, 2, 3, 72
Wyoming Valley Oral History Project, xvii, 147-51
Zaboroski, Simon, 87
Zakseski, Ed, 12-13
Zelonis, Herman, 3, 29, 35